How to Draw Dinosaurs

Volume 2

By Tracy Lee Ford

ISBN-13: 978-1535245234

ISBN-10: 1535245239

Table of Contents

Dedication
I dedicate this volume to my good friend Mike Fredericks.

I would like to thank Mike Fredericks for letting me write the How to Draw Dinosaurs articles for his magazine, Prehistoric Times. If it wasn't for him I wouldn't have been able to write and illustrate my articles. He is the man behind the magazine, the driving force, the producer and editor (though now he does have help) of the longest running Dinosaur/Fossil/Prehistoric magazine. I can't thank him enough for editing my articles to make them grammatically correct, and I'm sure I'm the cause of many of his grey hairs. In late 1995 (I think) I had started to buy Prehistoric Times and had bought the back issues at the time to fill out my collection (I'm happy to say I have a complete collection of Prehistoric Times). At the second Dinofest held in Arizona (1996) I had the pleasure of dinning with Mike. It was then that I brought up the idea of starting a series of articles on how to better depict dinosaurs by using their anatomy. He was up for it and shortly after that I started to write the articles. I have fun writing the articles and I'm glad to hear from some many people who use them. Luckily I still have lots of things to write about so the series will continue for quite some time.

Acknowledgements

I like to thank my family for their support for my passion for dinosaurs. George Olshevsky, whom has been my paleontological friend and mentor. Darren Tanke who, along with George, has helped me in getting my illustrations published. The paleontologists that have helped me over the years are: Jim Kirkland, and Ken Carpenter (who both helped me become a published 'paleontologist'), along with Tom Demere at the San Diego Natural History Museum, Bob McCord, Debbie Boaz and all the rest at the Mesa Southwest Museum (now called the Arizona Museum of Natural History), Dan Chure, Ralph Molnar, Peter Galton, Phil Currie, Jack Horner, John McIntosh, Spencer Lucas, Mike Brett-Surman, Don Glut, Greg Paul, Stephen Czerkas, Mark Hallett, Larry Martin, Mike Habib, Emily Buchholtz, Emily Brey, my bosses at Miners Gems and Minerals, Pat and Dana Dugan, Marel Dugan, Casey Dugan, and my fellow cohorts, Gerry Alverado, Jessica McConnell, Kendra Carty, Tonya Beach, Lisa Bolendar, and I apologize to all the others that I have forgotten.

Preface

For as long as I can remember I've been interested in dinosaurs. There is not one book, toy or movie that got me interested in dinosaurs. When I was a child I could find dinosaur toys or books no matter where my family went. When the Los Angeles County Library moved the books around, I found the dinosaur books, which were put on the top shelf. I liked to draw dinosaurs, lizards, birds, fish crabs, octopi, etc. But I never shaded, just drew the outlines. For the life of me, I don't know why I didn't shade the drawings.

When I graduated from High School I didn't know how to continue my interest. My High School kept trying to get men into Oceanography, even though I kept saying I wanted to study Paleontology. By luck, one day in 1978 (I had graduated in 1976), my family went to the Los Angeles Natural History Museum (we've moved to the San Diego area in 1970). In the bookstore was a small book written by George Olshevsky. It had all the dinosaur names known at that time, many of which I hadn't heard of. I wrote to his Toronto address but at first he did not reply. However months later George did write back, and it turned out he was living in San Diego, and not far from where I lived. I had asked him about some of the dinosaurs, and specifically about *Compsognathus corallestris*. I saw an illustration of it in a book and it had paddles for front legs. He said he didn't have time to go to libraries to look up the new dinosaurs. I told him I could do that for both of us, and photo copy the articles. Not only did I copy articles for him, but I also copied articles for paleontologists and mailed the articles to them.

This started a long friendship. In 1984 George sponsored me so I could join the SVP (Society of Vertebrate Paleontology). We went to the SVP meeting held at Berkeley. That is where I first met the paleontologists I had often read about. Since then I've missed a few SVP meetings, due to my work schedule and finances. George introduced me to many paleontologist and Dino artists. George started to publish Archosaurian Articulations, and he let me do the illustrations for it. I had to learn pen and ink, since I had only worked in pencil. My father introduced me to stippling, and I taught myself that art style. Darren Tanke also asked me if I would make some illustrations for him. My biggest break in 'Dino' illustrations was doing the majority of art for the *Dinosaur Society Dinosaur Encyclopedia,* and illustrating George's articles for Gakken Mook's *Dinosaur Frontline*. Thanks to George (to whom I am eternally grateful) and all the paleontologists I've met and had the pleasure of talking Dinos too, my once hobby is finally paying off, both scientifically and monetarily.

Why do a book or a series of articles on drawing dinosaurs? As I stated before, I have always been interested in dinosaurs. It wasn't until after I visited museums to see the mounted specimens that I found out what was actually known. To my dismay, I found out that many of the specimens I had thought were complete, weren't, and that some had been mounted incorrectly. After doing library research I started to find out what was known about dinosaur skeletons and how the bones fit together. Another big help was being able to go to symposiums and talk to paleontologists and fellow artist. It was through theses experiences that the dinosaurs, and for that matter, fossil animal life, came to life for me. This knowledge is what I want to pass on to artists and lay persons to help them to understand these wonderful animals better, and through their bones and how they articulate. There are a lot of good artists that don't know how the animals were, anatomically speaking, and this book will help them to correct their art form. Also, there are a lot of publishers who don't use artists who now their material, and this book will hopefully help them also. This is the first in a series of books. Each book will cover 25 articles (except for the first one, which will also cover an article I did for a different publication. In each volume I will had Editor: Notes that will update the subject of that article. Also, this volume I have changed some of the illustrations up to date them. Especially chapter two, in which I added dozens of ceratopian skulls to make the chapter more up to date.

Please visit my websites: http://www.dinohunter.info and http://www.paleofile.com

PREHISTORIC TIMES

April / May 2001   No. 47

The PT Interview: Artist Karen Carr

Prehistoric Terror Birds!

Don Lessem & The Biggest Dinosaur Ever!

Plus Dino Art, Articles & Reviews

U.S. $5.95 • Canada $6.95

6

Ford, T. L., 2001, How to Draw Dinosaurs. Head games, Part 1. Crests, frills, horns, and bumps, the heads of theropods...: Prehistoric Times, n. 47, p. 14-15.

Chapter 1
Part 1. Crests, frills, horns, and bumps, the heads of theropods...

As noted in the last installment of How to Draw Dinosaurs, this issue's topic is the heads of theropods and how to depict their wide variety of cranial displays; many of which are misinterpreted by artists. The width of theropods's skulls are quite deceiving and will also be discussed.

Coelophysis has a long, narrow skull with two small ridges running along the edges of the nasals (figure 1a). Some dispute this, but I have seen these small ridges on some specimens. *Syntarsus kayentakate* (a relative of *Coelophysis*), has two larger ridges or small 'frills' on its nasal (figure 1b). The frill starts behind the nasals and ends just in front of the orbits. *Dilophosaurus wetherill*, the now infamous 'spitting' theropod from Jurassic Park, has a long skull with two very large, very think 'frills' on the nasals which start behind the naries and end just behind then eyes (figure 1c). Unfortunately, Samuel Wells, who described the 2 known specimens did not describe the better preserved skull that had the frills. There are several new specimens mainly known form isolated elements. The premaxilla is slightly pinched giving the tip of the snout a tear dropped shape in dorsal and ventral views.

Abelisaurid's have some of the most bizarre looking heads of theropods. *Abelisaurus*, the namesake of the family, has a large, bulky skull that lacks horns but has rugosities (i.e. bumps and pits on the bone surfaces) on the nasals (this is true for all known abelisaurids, figure 1e). The skulls of abelisaurids are wider than other theropods. Bonaparte at one time believed the nasals had a horny sheath on them (I'm not sure he still believes this). Phil Currie has also mentioned this for other theropods (i.e. tyrannosaurids) that has rugosities on the nasals.

Carnotaurus sastre has a very short skull with two very wide, short round horns extending laterally over the orbits. These horns are not laterally compressed as some have depicted. The skull has been crushed laterally giving it a mistaken appearance (figure 1f). The recently described material of *Majungosaurus crenatissimus*, has a short laterally compressed skull with a single 'done' on the skull in-between the eyes (figure 1g).

Monolophosoaurus jiangi has a single 'crest 'on the top of the skull that terminates just behind the eyes. The skull is misleading in that the 'crest' makes the skull look thick, which it isn't; it is laterally compressed and thin (figure 2a).

Allosaurus, one of the more famous theropods, has two small ridges running along the nasals (these ridges are not smooth but have small 'bumps' on them), with two laterally compressed 'horn's in front of the eyes, not over the eyes. The 'horn's can be either rounded or slightly pointed (figure 2b). Other theropods with two ridges include, *Sinraptor, Acanthosaurus, Carcharodontosaurus*, and *Gigantosaurus*.

Some tyrannodaurids also have laterally compressed horn's in front of the eyes: *Albertosaurus* (figure 2c), *Gorgosaurus* (figured 2d), and *Daspletosaurus* (Figure 2 e). Tyrannosaurus rex (figure 2f) lacks these 'horn's, but does have large rugosities over the eyes. The nasals lack ridges but have rugositites (Figure 2b). In some Tyrannosaurus specimens the nasals are nearly smooth (Sue), while others are very rug (bumpy) (Figure 2h).

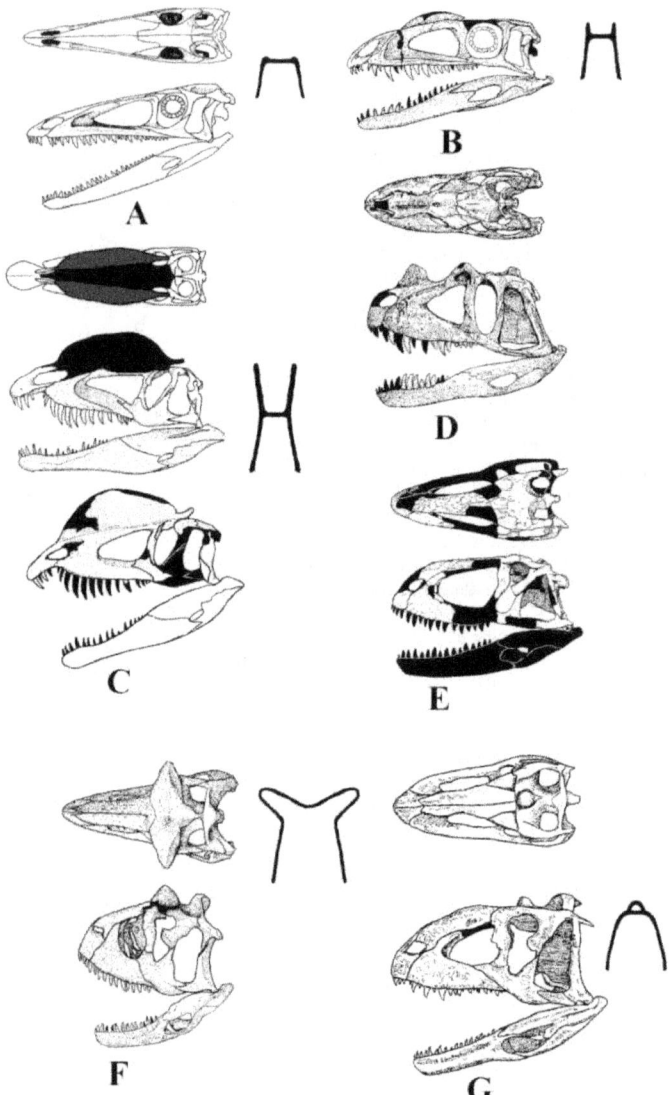

Figure 1) Skulls of theropods in lateral, dorsal, and cross-section: A) *Coelophysis bauri* after Colbert, 1989; B) *Syntarsus kayentakatae* after Rowe, 1989; C) *Dilophosaurus wetherilli*, top UCMP 37302, after Welles, 1984, bottom UCMP 77270, modified from Pickering, 1984; D) *Ceratosaurus nasicornis* after Gilmore, 1920; E) *Abelisaurus comahuensis* after Bonaparte & Novas, 1985; F) *Carnotaurus sastrei* after Bonaparte, Novas, & Coria, 1990; and G) *Majungasaurus crenatissimus* after Sampson & Witmer, 2007.

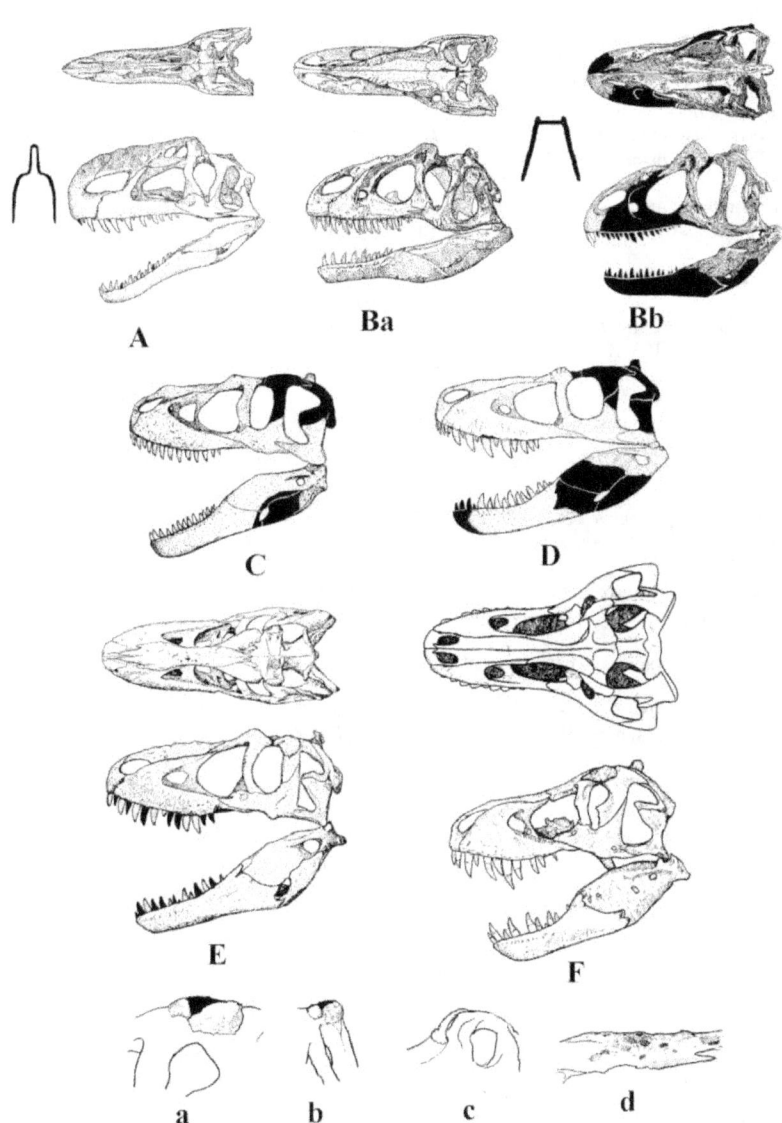

Figure 2) Skulls of theropods continued: A) *Monolophosaurus jiangi* after Zhao & Currie, 1994; B) *Allosaurus fragilis*; Ba) after Gilmore, 1920; Bb) after Madson, 1976; C) *Albertosaurus sarcophagus* modified from Osborn, 1905; D) *Gorgosaurus libratus* modified from Lambe, 1917; E) *Daspletosaurus torosus* after Russell, 1970; F) *Tyrannosaurus rex*, top LACM 23844, after Molnar, 1991, bottom after Larson, 2008; G) close up of individual skull bones of the *Tyrannosaurus rex* specimen, 'Stan' BHI 3033; a) lateral view of rugosities over the orbit; b) anterior view of rugosities over the orbit; c) dorsal view of rugosities over the orbit; and d) nasal region.

Bibliography

Bonaparte, J. F., and Novas, F. E., 1985, *Abelisaurus comahuensis*, n. g. n. sp., Carnosauria del cretacico Tardio de Patagonia: Ameghiniana, v. 21, n. 2-4, p. 259-265.

Bonaparte, J. F., Novas, F. E., and Coria, R. A., 1990, *Carnotaurus sastrei* BONAPARTE, the Horned, Lightly Built Carnosaur from the Middle Cretaceous of Patagonia: Contributions in Science, n. 416, p. 1-41.

Colbert, E. H., 1989, The Triassic Dinosaur *Coelophysis*: Museum of Northern Arizona Bulletin n. 57, 160pp.

Russell D. A., 1970, Tyrannosaurs from the Late Cretaceous of Western Canada: National Museum of Natural Science, Publication in Paleontology, n. 1, p. 1-34.

Gilmore, C. W., 1920, Osteology of the Carnivorous Dinosauria in the United States National Museum, with special reference to the genera *Antrodemus* (*Allosaurus*) and *Ceratosaurus*: Bulletin of the United States National Museum, n. 110, p. 1-159.

Lambe, L. M., 1917, The Cretaceous theropodous dinosaur Gorgosaurus: Geological Survey of Canada Memoir, v. 100, p. 1-84.

Larson, P., 2008, Variation and sexual dimorphism in *Tyrannosaurus rex*: In: *Tyrannosaurus rex*, the tyrant king, edited by Larson, P., and Carpenter, K., Indiana University Press, p. 103-128.

Madsen, J. H. jr., 1976, *Allosaurus fragilis* a revised osteology: Utah Geological and Mineral Survey, Bulletin, n. 109, p. 1-163.

Molnar, R., E., 1991, The Cranial morphology of *Tyrannosaurus rex*: Palaeontographica Abt. A, v. 217, lfg. 4-6, p. 137-176.

Osborn, H. F., 1905, Tyrannosaurus and other Cretaceous carnivores Dinosaurs: Bulletin of the American Museum of Natural History, v. 21, p. 259-265.

Pickering, S., 1994, An extract from Archosauromorpha: Cladistics & Osteologies, (*Dilophosaurus*), 70pp.

Rowe, T., 1989, A new species of the theropod dinosaur *Syntarsus* from the Early Jurassic Kayenta Formation of Arizona: Journal of Vertebrate Paleontology, v. 9, n. 2, p. 125-136.

Sampson, S. D., and Witmer, L. M., 2007, Craniofacial anatomy of *Majungasaurus crenatissimus* (Theropoda: Abelisauridae) from the Late Cretaceous of Madagascar: Journal of Vertebrate Paleontology, v. 27, supplement to n. 2, Memoir 8, p. 32-102.

Welles, S. P., 1984, *Dilophosaurus wetherilli* (Dinosauria, Theropoda) osteology and comparisons: Palaeontographica Abt. A 185, lfg. 4-6, p. 85-180.

Zhao, X.-J., and Currie, P. J., 1994, A large crested theropod from the Jurassic of Xinjiang, People's Republic of China: In: Results from the Sino-Canadian Dinosaur Project. Canadian Journal of Earth Sciences, v. 30, p. 2027-2036.

For the Dinosaur Enthusiast and Collector

PreHistoric Times

June/July 2001 No. 48

The PT Interview:
Paleontologist
Scott Sampson

Burgess Shale

History of Dinosaur
Art & Sculpture

Giganotosaurus

and more!

U.S. $5.95 • Canada $6.95

©R.DELGADO
02-01

Ford, T. L., 2001, How to Draw Dinosaurs; The ceratopians, the horned frilled dinosaurs: Prehistoric Times, n. 48, p. 14.

The Ceratopians

By Tracy L. Ford

The most distinctive part of the ceratopians anatomy is their ornate horned skulls. The ornamentations vary from being a boss (a lump of reugose bone) over the nose to a long laterally compressed horn. The back of the skull, the frill, can be highly ornamented (with spikes or epiposiptials; small round to pointed laterally compressed horns running along the edge of the frill) to nearly plane. This article will go over some of the differences and similarities. I will not be going into the Protoceratopians, since they lack any major ornamentation on the head.

The ceratopians are divided into two major groups, the Centrosaurians and Chasmosaurians. The ceratopians can be broken down into subfamilies (and can be broken down even further in tribes, which I won't go into). Their skull isn't wide as is typically depicted but narrow. They all have a rostral bone (rostrum) or what has been commonly called a beak like a parrot's. The beak differs from a parrot's in that it was fused to the premaxilla. A parrot's beak is kinetic (i.e. can move independently from the skull) giving it its strong bite. I believe that the most potentially lethal aspect of the skull was its beak not its horns.

The skull, itself, can be long or short. The jugal is bone below the eye and the epijual is the part of the jugal that sticks out below the eyes. The squamosal is behind the eyes and has a 'U'-shaped edge behind the jugal. The Quadratojual is in between the jugal and squamosal and has the joint that connects the skull to the lower jaw. The 'U'-shaped part of the squamosal is where the ear would have been (Figure 1). The lower jaw (Dentary) has the front part of the beak, the premaxilla.

The squamosal can be short (as in Centrosaurians and some Chasmosaurians; i.e. some *Triceratops* species) or be long, as in most Chasmosaurians. The parietal extends behind the orbital horns to the back of skull. The middle section (sometimes called a bar) can be straight, wavy or as in the Pipestone Creek (about 30 miles due west of Edmonton, Alberta, Canada) *Pachyrhinosaurus* have horns on the parietal (more towards the orbit). They have been found with 1 to 3 horns on the parietal. The squamosal and parietal make up the frill. The epiposiptials run along the edge of the squamosal and parietal. They can be small rounded bones to large horns (*Styracosaurus*, *Pachyrhinosaurus* etc). For the most part there are 9 epiposiptials per side.

The frill in most ceratopians has two holes (fenestera). *Triceratops* and *Aviceratops* has a solid frill. In Chasmosaurs the frill is vary large and can extend very far past where the skull connects to the neck. The frill can also be narrow to very wide.

The nasal horn can be a boss (*Pachyrhinosaurus*, *Achelousaurus*), or a horn. It can be small (*Chasmosaurus*, *Triceratops*, *Torosaurus*) or very large (*Centrosaurus* and *Styracosaurus*). They can be straight (*Centrosaurus* and *Styracosaurus*), curve forward (in some Centrosaurs and *Einiosaurus*) or slightly backwards (in some Centrosaurs). The orbit horns can range from curving forward to curving backward (*Triceratops* has a very large range). New discoveries will show that the variety of horns in ceratopians is even greater than it now known.

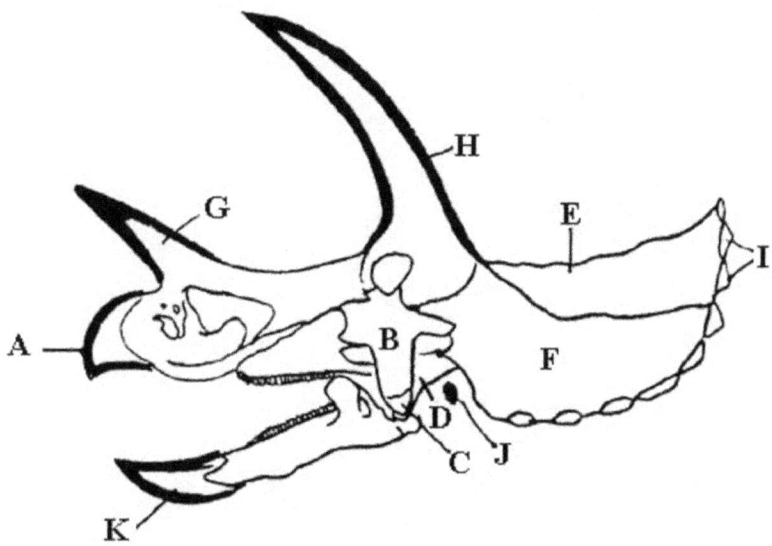

Figure 1). Skull of *Triceratops serratus*; A) rostrum; B) jugal; C) epijugal; D) quadratojugal; E) pariatal; F) squamosal; G) nasal horn; H) orbital horns; I) epiposiptials; J) ear; K) predentary.

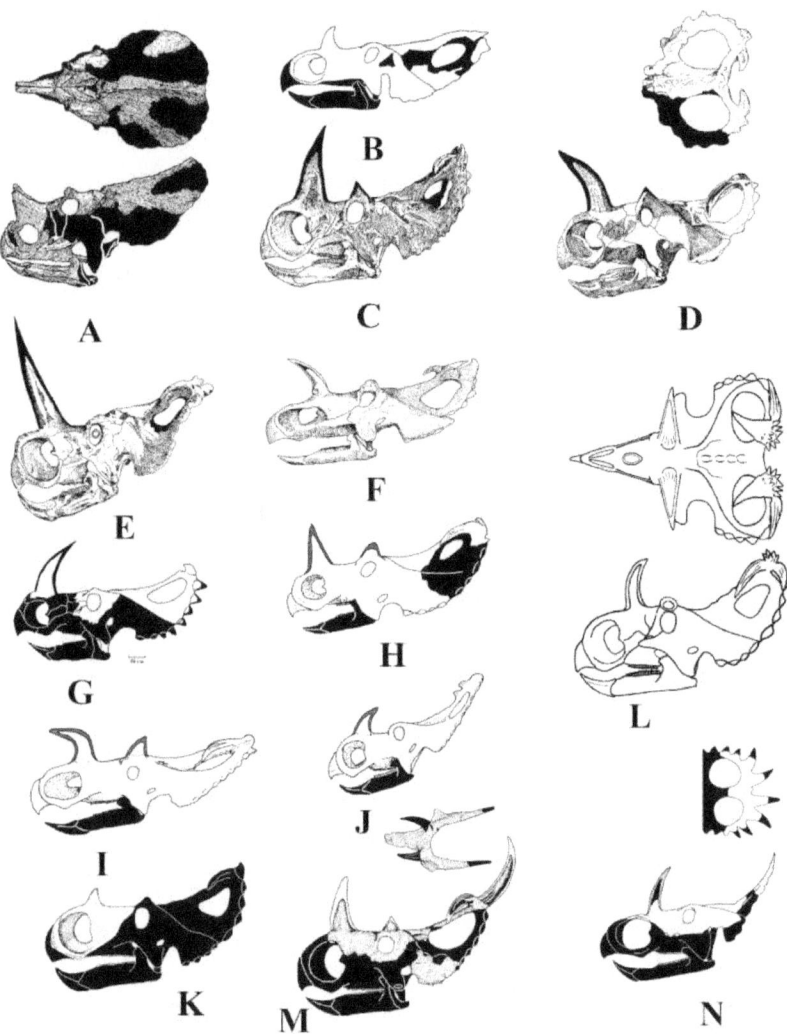

Figure 2). Skulls of centrosaurines (not to scale); A) *Brachyceratops montanensis* (USNM 7951) after Gilmore, 1914; B) *Monoclonius? lowei* after Sternberg, 1940 (NMC 8790); C-K) *Centrosaurus apterus* C) *Centrosaurus apterus* (ROM) after Lull, 1933; D) = *Centrosaurus flexus*, AMNH 5239) after Lull, 1933, dorsal view of frill and lateral view; E) = *Monoclonius nasicornis*, AMNH 5351) after Brown, 1917; F) = *Centrosaurus longirostris*, NMC 8795) after Sternberg, 1940; G) = *Monoclonius dawsoni* NMC 1173) Lambe, 1902 ; H-I) after Frederickson, et al., 2014; H) (RTMP 1994.182.0001); I) (UALVP 11735) J) (RTMP 1992.082.0001); K) (RTMP) from Darren Tanke; L) *Coronosaurus brinkmani* = *Centrosaurus brinkmani* RTMP 2002.68.1) modified from Ryan, et al, 2012, Ryan & Russell, 2005, in dorsal and lateral views.; M) *Spinops sternbergorum* (NHMUK R16307) modified from Farke, et al., 2011, dorsal view, and lateral views; and N) *Sinoceratops zhuchengensis* (ZCDM V0010) modified from Xu, et al.,   2010.

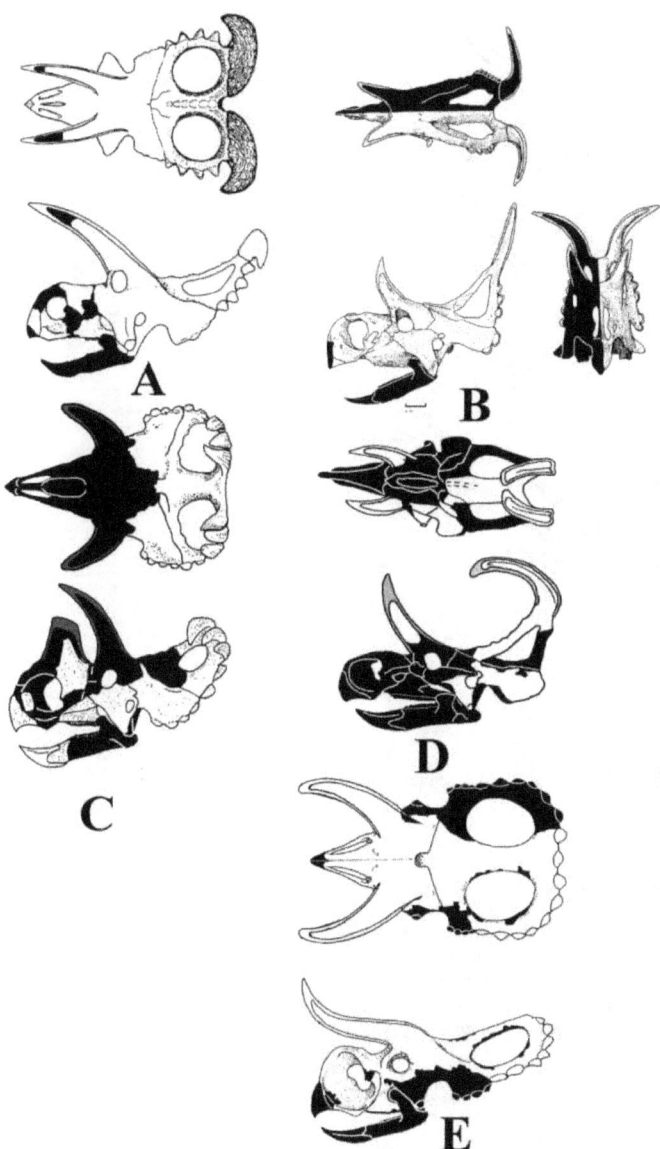

Figure 3). Skulls of centrosaurines (not to scale): *Albertaceratops nesmoi* (composite skull) modified from Ryan, 2007, dorsal and lateral views; B) *Diabloceratops eatoni* (UMNH VP 16699) modified from Kirkland & Deblieux, 2010, dorsal lateral and anterior views; C) *Wendiceratops pinhornensis* (RTMP 2011.051.0009) after Evans & Ryan, 2015, dorsal and lateral views; D) *Machairoceratops cronusi* (UMNH VP 20550) modified from Lund, et al, 2016, dorsal and lateral views; and E) *Nasutoceratops titusi* (UMNH VP 16800) after Sampson, et al., 2013.

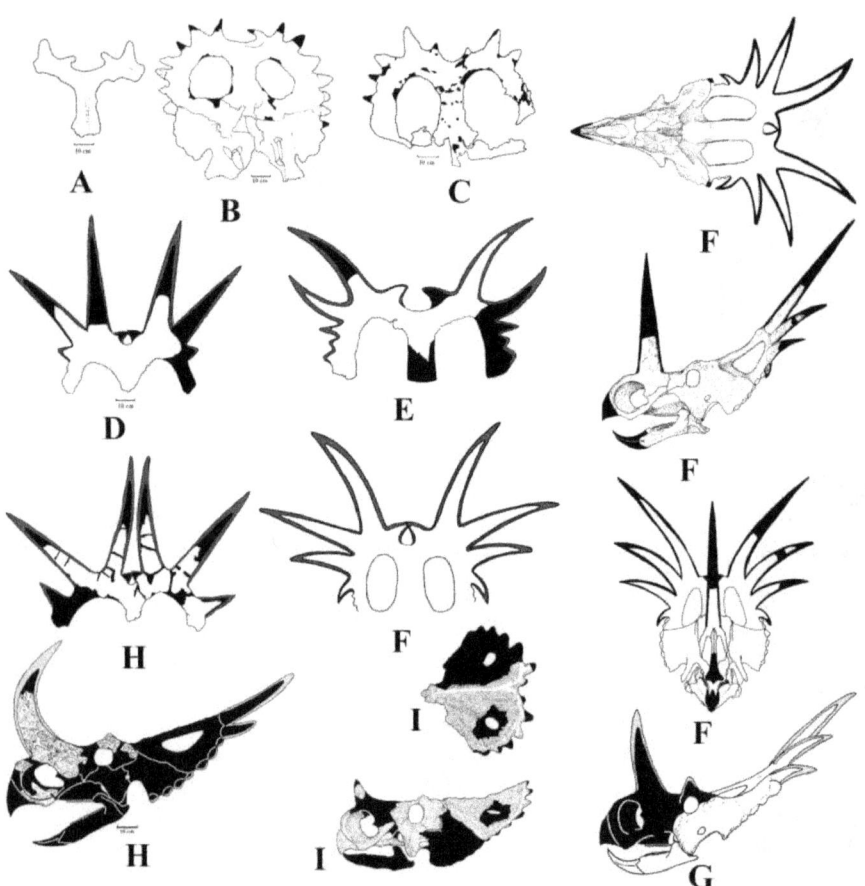

Figure 4). Skulls of centrosaurines/styracosaurians (not to scale): A-G) *Styracosaurus*; A-F) *Styracosaurus albertensis*; A) juvenile parietal, (RTMP 99.55.2), after Ryan, et al., 2007; B) sub adult (RTMP 98.126.1), after Ryan, et al., 2007; C) subadult (RTMP 89.97.1), after Ryan, et al., 2007; D) subadult (RTMP 88.36.20), after Ryan, et al., 2007; E) subadult (ROM 1436), after Ryan, et al., 2007; F) composite new reconstruction, frill, dorsal lateral and anterior views; G) *Styracosaurus albertensis* (= *Styracosaurus parksi*, AMNH 5372) modified from Brown, & Schlaikjer, 1937; H-I) *Rubeosaurus ovatus* (= *Styracosaurus ovatus*); H) *Rubeosaurus ovatus* (USNM 11869), modified from McDonald & Horner, 2010; and I) *Rubeosaurus ovatus* (referred *Brachyceratops montanensis*, USNM 14765), modified from Gilmore, 1939.

16

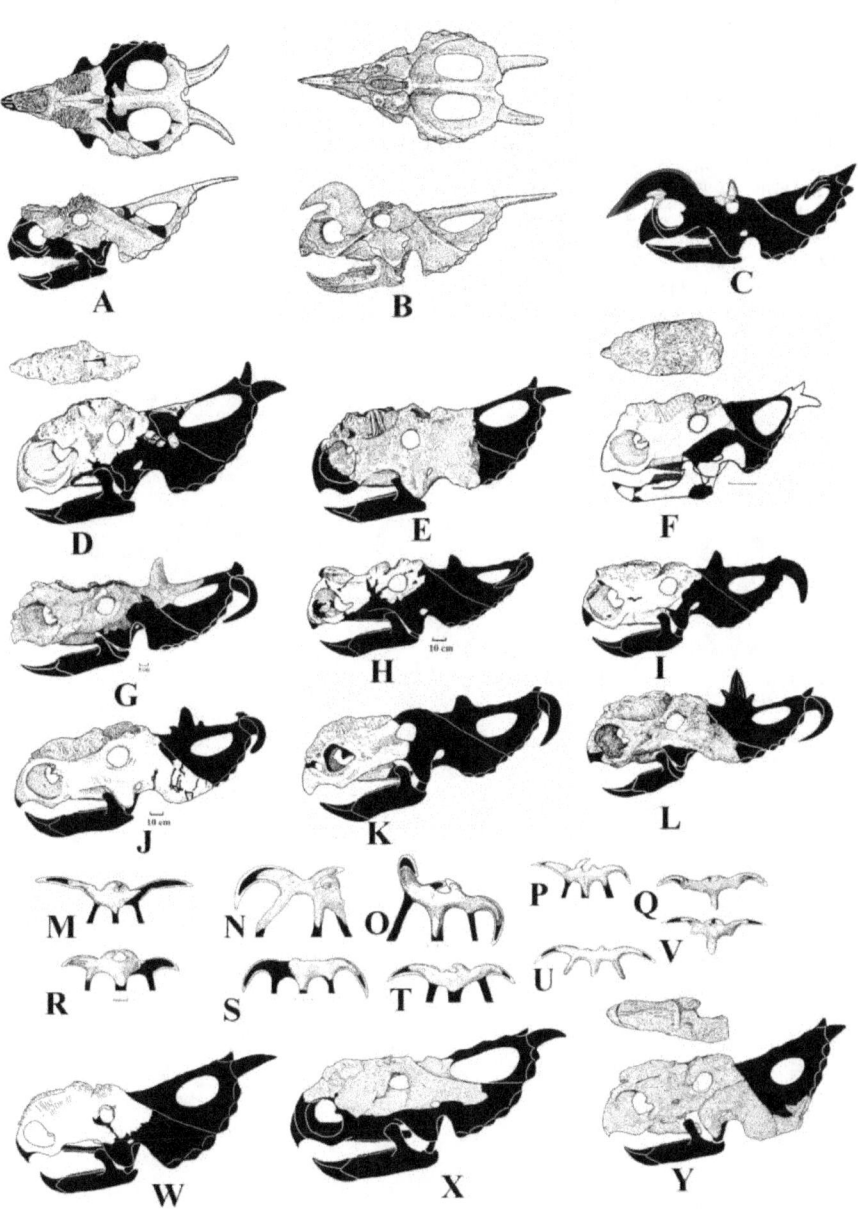

Figure 5). Skulls of centrosaurines/pachyrhinosaurians (not to scale): A) *Achelousaurus horneri* (MOR 485) after Sampson, 1995, dorsal and lateral view; B) *Einosaurus procurvicornis (*composite skull) after Sampson, 1995, dorsal and lateral views; C) *Centrosaurus? recurvicornis* (AMNH 3999) modified from Cope, 1890; D) D-Y) *Pachyrhinosaurus*; D-F) *Pachyrhinosaurus canadensis*; D) (Drumheller and District Museum Society Museum) modified from Langston, 1967, dorsal and lateral views; E) (NMC 8867) modified from Sternberg; F) (NMC 8860) modified from Langston, 1976, dorsal and lateral views; G-V) *Pachyrhinosaurus lakustai*; G) (RTMP 1986.55.258), after Currie, et al., 2008;) H) (RTMP 1987.55.320), after Currie, et al., 2008; I) (RTMP 1987.55.156), after Currie, et al., 2008; J) (RTMP 1989.55.188), after Currie, et al., 2008; K) (RTMP 1987.55.285), after Currie, et al., 2008; L) (RTMP 1989.55.1234), after Currie, et al., 2008; M-V) parietal frills of *Pachyrhinosaurus lakustai*; M) (RTMP 1986.55.258), after Currie, et al., 2008; M) (RTMP 1986.55.239); N) (RTMP 1987.55.210) pathologic frill; O) (RTMP 1989.55.1085); P) (RTMP 1989.55.1503); Q) (RTMP 1987.55.141); R) (RTMP 1986.55.261); S) (RTMP 1986.55.113); T) (RTMP 1988.55.46); U) (RTMP 1989.55.1144); V) (RTMP 1989.55.1241); W) *Pachyrhinosaurus perotorum* (DMNH 21200), modified from Fiorillo, & Tykoski, 2012; X) *Pachyrhinosaurus* nova? (UCMP, from a photograph taken in 1991); and Y) *Pachyrhinosaurus* nova (RTMP 2002.76.1) after Ryan, et al., 2010.

18

Figure 6). Skulls of chasmosaurines (not to scale); A) *Anchiceratops ornatus* (AMNH 5251) modified from Brown, 1914; B) *Anchiceratops ornatus* (AMNH 5259), after Brown, 1914, dorsal and lateral views; C) *Anchiceratops longirostris* (NMC 8535), after Sternberg, 1929, dorsal and lateral views; D) *Arrhinoceratops brachyops* (ROM 796) modified from Parks, 1925, dorsal and lateral views; E) *Coahuilaceratops magnacuerna* (CPC 276) modified from Loewen, et al., 2010; F) *Kosmoceratops richardsoni* (UMNH VP 17000), after Sampson, et al., 2010, dorsal and lateral views; G) *Xenoceratops foremostensis* (CMN 53282) modified from Ryan, et al., 2012, dorsal and lateral views; H) *Ojoceratops fowleri (*SMP VP-1865) after Sullivan & Lucas, 2010; I) *Spiclypeus shipporum* (CMN 58071), after Mallon, et al., 2016, dorsal and lateral views; K) *Regaliceratops peterhewsi* (RTMP 2005.055.0001), after Brown & Henderson, 2015, dorsal and lateral views.

Figure
7). Skulls of chasmosaurines (not to scale); A-E) *Agujaceratops mariscalensis* (= *Chasmosaurus mariscalensis*); A-D) modified from Lehman, 1989; A) (UTEP P. 37.7.086, and 046, 065, and 066); B) (UTEP 062, 064, 073 082); C) (UTEP 045, 087, 088, 091); D) UTEP 059, 072, 081, 085, and 095); E) (lower TMM 43098-1, after Forster, et al., 1993, and upper based on 046m 063, 065, 070, and 071, modified from Lehman, 1989); F) *Bravoceratops polyphemus* (TMM 46015-1) after Wicks, & Lehman, 2013; H-J) *Pentaceratops sternbergi:* H) (AMNH 1624, MNA PL. 1747, and UNM FKKK-081) after Lehman, 1993, front, dorsal and lateral views; I) (MNA PL. 1747) modified from Lehman, 1993; J) = *Titanoceratops ourano*s (OMNH 10165), modified from Longrich, 2011; K) *Pentaceratops fenestratus* (PMU R200) modified from Wiman, 1930; L) *Pentaceratops aquilonius* (SDNHM 43470) modified from Longrich, 2014; M-P) *Torosaurus latus;* M) ANSP 15192; N) = *Torosaurus gladius (*YPM 1831) modified from Marsh, 1891, dorsal and lateral views; O) (YPM 1830), modified from Farke, 2007, dorsal and lateral views; P) *Torosaurus?* *utahensis* (USNM 15583) modified from Gilmore, 1946; K) *Torosaurus latus?* = *Triceratops* sp (AMNH 970) modified from Lull, 1933.

8). Skulls of chasmosaurines (not to scale); A-D) *Chasmosaurus belli*; A) dorsal (NMC 0491) after Lambe, 1902; B) (ROM 843) after Godfrey & Holmes, 1995, dorsal and lateral views; C) (NMC 2245) after Godfrey & Holmes, 1995, dorsal and lateral views; D) = *Chasmosaurus brevirostris (ROM 839) after Lull,* 1933; E-F) *Chasmosaurus russelli*; E) (NMC 2280) after Godfrey & Holmes, 1995: F) (RTMP 81.19.175) after Godfrey & Holmes, 1995; G) *Utahceratops gettyi* (UMNH VP 16784), after Sampson, et al., 2010, dorsal and lateral views; H) *Judiceratops tigris* (YPM YUPPU 022404), modified from Longrich, 2013, dorsal and lateral views; I) *Medusaceratops lokii (*right WDC-DJR-002, left WDC-DJR-001) after Ryan, et al., 2010, dorsal view; J-M) *Eoceratops canadensis*; J) (NMC 1254) after Lambe, 1915; K) = *Mojoceratops perifania* (RTMP 1983.25.1) after Longrich, 2010, dorsal and lateral views; L) = *Chasmosaurus kaiseni* (AMNH 5401), after Brown, 1933; M) (UAVP 40) after Tyson, 1977, 1981; N) *Vagaceratops irviensis* (NMC 41357) modified from Holmes, et al., 2001, dorsal, lateral and anterior views; O) *Mercuriceratops gemini (*ROM 64222), modified from Ryan, et al., 2014; and P) *Zuniceratops christopheri* (composite skull), modified form Wolfe, et al., 2010.

Figure 9). Skulls of triceratopians (to scale). A) *Nedoceratops hatcheri* (= *Diceratops hatcheri* Hatcher vide Lull, 1905/Marsh, Hatcher & Lull, 1907,= *Diceratus hatcheri* (Hatcher vide Lull, 1905/Marsh, Hatcher & Lull, 1907) Mateus, 2008) Ukrainsky, 2007 (USNM 2412) modified from Hatcher, et al., 1905, dorsal and lateral views; B) *Nedoceratops? albertensis* (= *Triceratops albertensis* Sternberg, 1949, NMC 8862) modified from Sternberg, 1949; C) *Avaceratops lammersorum* (ANSP 15800) modified from Dodson, 1986; D) *Avaceratops lammersorum* (MOR 692) modified from Penkalski, 1994; E) *Tatankaceratops sacrisonorum* (BHI 6226) modified from Ott, & Larson, 2010; F) *Eotriceratops xerinsularis* (RTMP 2002.57.7) modified from Wu, et al., 2007; G) *Triceratops? eurycephalus* (MCZ 1102), modified from Schlaikjer, 1935; H-M, O-Q) *Triceratops horridus*; H) = *Triceratops serratus* (YPM 1823) modified from Marsh, 1890, dorsal and lateral views; I) (YPM 1820) modified from Hatcher, vide Lulll, 1905/Marsh, Hatcher & Lull, 1907; J) = *Triceratops flabellatus* (YPM 1821) after Lull, 1934; K) = *Triceratops obtusus* (USNM 4720) modified from Lull, 1933; L) = *Triceratops elatus* (USNM 1201) after Marsh, 1891; M) = *Triceratops calicornis* (MNHN 1912.20) modified from Marsh, 1989; N) *Triceratops? alticornis* = *Bison alticornis* (USNM 1817E) modified from Marsh, 1887; O) (TCM 2001.93.1) after Glut, 2006; P) (UCMP 113697), after Forster, 1996; Q) (BYU) from a photograph I took in the 1990's; R-Y) *Triceratops prorsus*; R) (YPM 1822) modified from Hatcher vide Lull, 1905/Marsh, Hatcher, & Lull, 1907, dorsal and lateral views; S) (USNM 2100) modified from Hatcher vide Lull, 1905/Marsh, Hatcher, & Lull, 1907, dorsal and lateral views; T) = *Triceratops brevicornus* (BSP 1964 I 458) modified from Hatcher vide Lull, 1905/Marsh, Hatcher, & Lull, 1907, dorsal and lateral views; U) (SDSM 2760) after Foster, 1996; V) = *Triceratops cf. brevicornis* (Mineralogisch-Geologisch Museum) after Schuyf, 1969; W) (SMNH P1163.4) modified from Tokaryk, 1986; X) (EM P15.1) after Tokaryk, 1986; and Y) (UND 3000) after Holland, 1997.

Bibliography

Brown, B. B., 1914, *Anchiceratops*, a new genus of horned dinosaurs from the Edmonton Cretaceous of Alberta. With discussion of the origin of the Ceratopsian crest and the brain casts of *Anchiceratops* and *Trachodon*: Bulletin of the American Museum of Natural History, v. 33, p. 539-548.

Brown, B. B., 1914, A complete skull of *Monoclonius*, from the Belly River Cretaceous of Alberta: Bulletin of the American Museum of Natural History, v. 33, p. 549-558.

Brown, B. B., 1917, A complete skeleton on the horned dinosaur *Monoclonius*, and description of a second skeleton showing skin impressions: Bulletin of the American Museum of Natural History, v. 37, p. 281-306.

Brown, B. B., 1933, A new longhorned Belly River Ceratopsian: American Museum Novitates, n. 669, p. 1-3.

Brown, B. B., and Schlaikjer, E. M., 1937, The skeleton of *Styracosaurus* with the description of a new species: American Museum Novitates, n. 955, p. 1-12.

Brown, C. M., and Henderson, D. M., 2015, A new horned dinosaur reveals convergent evolution in cranial ornamentation in Ceratopsidae: Current Biology, v. 25, p. 1-8.

Currie, P. J., Langston, W. Jr., and Tanke, D. H., 2007, A new pachyrhinosaur from the Wapiti Formation of Grande Prairie, Alberta: In: Ceratopsian Symposium, Short Papers, Abstracts, and Programs, complied by Braman, D. R., p. 22-23.

Currie, P. J., Langston, W., jr., and Tanke, D. H., 2008, A new species of *Pachyrhinosaurus* (Dinosauria, Ceratopsidae) from the Upper Cretaceous of Alberta, Canada: In: A new horned dinosaur from an Upper Cretaceous Bone bed in Alberta: A publication of the National Research Council of Canada Monograph Publishing Progam, 1-108.

Danis, J., 1986, 7. Quarries of Dinosaur Provincial Park: In: Dinosaur Systematics Symposium, Field Trip Guidebook to Dinosaur Provincial Park, 2 June 1986, p. 43-51.

Diem, S., and Archibald, J. D., 2005, Range extension of southern chasmosaurine ceratopsian dinosaurs into northwestern Colorado: Journal of Paleontology, v. 79, n. 2, p. 251-258.

Dodson, P., 1986, *Avaceratops lammersi*: A New Ceratopsid from the Judith River Formation of Montana: Proceedings of the Academy of Natural Sciences of Philadelphia, v. 138, n. 2, p. 305-317.

Evans, D. C., and Ryan, M. J., 2015, Cranial anatomy of *Wendiceratops pinhornensis* gen. et sp. nov., a Centrosaurine Ceratopsid (Dinosauria: Ornithischia) from the Oldman Formation (Campanian), Alberta, Canada, and the evolution of Ceratopsid nasal ornamentation: Public Library of Science (PLOS), One, v. 10, n. 7, 31 pp.

Farke, A. A., 2006, Cranial osteology and phylogenetic relationships of the chasmosaurine ceratopsid Torosaurus latus: In: Horns and Beaks, edited by Carpenter, K., Indiana University Press, p. 235-257.

Farke, A. A., Ryan, M. J., Barrett, P. M., Tanke, D. H., Braman, D. R., Loewen, M. A., and Grahm, M. R., 2011, A new centrosaurine from the Late Cretaceous of Alberta, Canada, and the evolution of parietal ornamentation in horned dinosaurs: Acta Palaeontologica Polonica, v. 56, n. 4, p. 691-702.

Fiorillo, A. R., and Tykoski, R. S., 2012, A new Maastrichtian species of the centrosaurine ceratopsid *Pachyrhinosaurus* from the North Slope of Alaska: Acta Palaeontological Polonica, v. 57, n. 3, p. 561-573.

Forster, C. A., Sereno P. C., Evans T. W., and Rowe T., 1993, A complete skull of *Chasmosaurus mariscalensis* (Dinosauria: Ceratopsidae) from the Aguja Formation (late Campanian) of West Texas: Journal of Vertebrate Paleontology, v. 13, n. 2, p. 161-170.

Frederickson, J. A., and Tumarkin-Deratzian, A. R., 2014, Craniofacial ontogeny in *Centrosaurus apertus*: PeerJ 2x252; DOI 1-.7717;peerja.252, 32pp

Gilmore, C. W., 1914, A New Ceratopsian Dinosaur from the Upper Cretaceous of Montana, with Note on *Hypacrosaurus*: Smithsonian Miscellaneous Collections, v. 63, n. 3, p. 1-10.

Gilmore, C. W., 1916, Vertebrate faunas of the Ojo Alamo, Kirtland, and Frutitland Formations: United States Geological Survey, Professional Papers, n. 18, p. 279-308.

Gilmore, C. W., 1917, *Brachyceratops*, a ceratopsian dinosaur from the Two Medicine Formation of Montana, with notes on associated fossil reptiles: United States Geological Survey professional paper, v. 103, p. 1-45.

Gilmore, C. W., 1920, Reptile reconstruction's in the United States National Museum: Annual Report of the Smithsonian Institution, 1918, p. 271-280.

Gilmore, C. W., 1920, The Horned Dinosaurs: Annual Report Smithsonian, Institution, 1920, p. 381-387.

Gilmore, C. W., 1939, Ceratopsian Dinosaurs from the Two Medicine Formation, Upper Cretaceous of Montana: Proceedings of the United States National Museum. 87, p. 1-18.

Gilmore, C. W., 1946, Reptilian fauna of the North Horn Formation of central Utah: United States Geological Survey professional paper, n. 210-C, p. 29-53.

Glut, D. F., 2006, Dinosaurs, the Encyclopedia, Supplement 4: McFarland & Company, Inc, 749pp.

Godfrey, S. J., and Holmes, R., 1995, Cranial morphology and systematics of *Chasmosaurus* (Dinosauria: Ceratopsidae) from the Upper Cretaceous of Western Canada: Journal of Vertebrate Paleontology, v. 15, n. 4, p. 726-742.

Hatcher, J. B., Marsh, O. C., and Lull, R. S., 1907, The Ceratopsia: Monographs of the United States Geological Survey, v. 49, p. 1-300.

Holmes, R., Forster, C. A., Ryan, M., and Shepherd, K. M., 2001, A new species of *Chasmosaurus* (Dinosauria: Ceratopsia) from the Dinosaur Park Formation of southern Alberta: Canadian Journal of Earth Sciences, v. 38, n. 1423-1438.

Holmes, R. B., Ryan, M. J., and Murray, A. M., 2006, Photographic atlas of the postcranial skeleton of the type specimen of *Styracosaurus albertensis* with additional isolated cranial elements from Alberta: Syllogeus, n. 75, 75pp.

Kirkland, J. I., and Deblieux, D. D., 2010, New basal centrosaurine ceratopsian skulls from the Wahweap Formation (Middle Campanian), Grand Staircase-Escalante National Monument, southern Utah: In: New Perspectives on Horned Dinosaurs. The Royal Tyrrell Museum Ceratopsian Symposium, edited by Ryan, M. J., Chinnery-Allgeier, B. J., and Eberth, D. A., Indiana University Press, Part Two, p. 99-116.

Lambe, L. M., 1914, On the fore-limb of Carnivorous dinosaur from the Belly River Formation of Alberta, and a new genus of Ceratopsia from the same horizon, with remarks on the integument of some Cretaceous Herbivorous Dinosaurs: The Ottawa Naturalist, v. 27, n. 10, p. 129-133.

Lambe, L. M., 1915, On *Eoceratops canadensis*, gen. nov., with remarks on other genera of Cretaceous Horned Dinosaurs: Canada Geological Survey Museum Bulletin, v. 12, Geological series, n. 24, p. 1-25.

Langston, W. Jr., 1967, The thick-headed ceratopsian dinosaur *Pachyrhinosaurus* (Reptilia Ornithischia). From the Edmonton Formation near Drumheller, Canada: Canadian Journal of Earth Sciences, v. 4, p. 171-186.

Langston, W. Jr., 1968, A further note on *Pachyrhinosaurus* (Reptilia: Ceratopsia): Journal of Paleontology, v. 42, n. 5, p. 1303-1304.

Langston, W. Jr., 1975, The ceratopsian dinosaurs and associated lower vertebrates from the St. Mary River Formation (Maastrichtian) at Scabby Butte, Southern Alberta: Canadian Journal of Earth Sciences, v. 12, p. 1576-1608.

Lehman, T. M., 1993, New data on the Ceratopsian dinosaur *Pentaceratops sternbergii* OSBORN from New Mexico: Journal of Paleontology, v. 67, n. 2, p. 279-288.

Loewen, M. A., Sampson, S. D., Lund, E. K., Farke, A. A., Aguillon-Martinez, M. C., de Leon, C. A., Rodriguez-de la Rosa, R. A., Getty, M. A., and Eberth, D. A., 2010, Horned dinosaur (Ornithischia: Ceratopsidae) from the Upper Cretaceous (Campanian) Cerro del Pueblo Formation, Coahuila, Mexico: In: New Perspectives on Horned Dinosaurs. The Royal Tyrrell Museum Ceratopsian Symposium, edited by Ryan, M. J., Chinnery-Allgeier, B. J., and Eberth, D. A., Indiana University Press, Part Two, p. 99-116.

Longrich, N., 2010, *Mojoceratops perifania*, a new chasmosaurine ceratopsid from the Late Campanian of western

Canada: Journal of Paleontology, v. 84, n. 4, p. 681-694.

Longrich, N. R., 2011, *Titanoceratops ouranos*, a giant horned dinosaur from the late Campanian of New Mexico: Cretaceous Research, v. 32, p. 264-276.

Longrich, N. R., 2013, *Judiceratops tigris*, a new horned dinosaur from the Middle Campanian Judith River Formation of Montana: Bulletin of the Peabody Museum of Natural History, v. 54, n. 1, p. 51-65.

Longrich, N. R., 2014, The horned dinosaurs *Pentaceratops* and *Kosmoceratops* from the upper Campanian of Alberta and implications for dionsaur biogeography: Creaceous Research, v. 51, p. 393-308.

Lucas, S. G., Sullivan, R. M., and Hunt, A. P., 2006, Re-Evaluation of *Pentaceratops* and *Chasmosaurus* (Ornithischia: Ceratopsidae) in the Upper Cretaceous of the western interior: In: Late Cretaceous Vertebrates from the Western Interior, edited by Lucas, S. G., and Sullivan, R. M., New Mexico Museum of Natural History & Science, Bulletin 35, p. 367-370.

Lull, R. S., 1903, Skull of *Triceratops serratus*: Bulletin of the American Museum of Natural History, v. 19, p. 685-695.

Lull, R. S., 1905, Restoration of the Horned dinosaur *Diceratops*: American Journal of Science, 4th series, v. 20, p. 420-422.

Lull, R. S., 1933, A revision of the Ceratopsia or Horned Dinosaurs: Memoirs of the Peabody Museum of Natural History, v. 3, part 3, p. 1-175.

Lull, R. S., 1934, Skull of *Triceratops flabellatus* recently mounted at Yale: The American Journal of Science, 5th series, v. 28, p. 439-442.

Lund, E. K., O'Connor, P. M., Loewen, M. A., and Jinnah, Z. A., 2016, A new centrosaurine ceratopsid, *Machairoceratops cronusi* gen et sp. nov., from the Upper Sand Member of the Wahweap Formation (Middle Campanian), Southern Utah: Public Library of Science (PLOS), One, v. 11, n. 5, 21 pp.

Lund, E. K., Sampson, S. D., and Loewen, M. A., 2016, *Nasutoceratops titusi* (Ornithischia, Ceratopsidae), a basal Centrosaurine Ceratopsid from the Kaiparowits Formation, southern Utah: Journal of Vertebrate Paleontology, v. 36, n. 2, e1054936. 26 pp.

Mallon, J. C., Ott, C. J., Larson, P. L., Iuliano, E. M., and Evens, D. C., 2016, *Spiclypeus shipporum* gen. et sp. nov., a boldly audacious new chasmosaurine ceratopsid (Dinosauria: Ornithischia) from the Judith River Formation (Upper Cretaceous: Campanian) of Montana, USA: Public Library of Science (PLOS), One, v. 11, n. 5, 40 pp.

Marsh, O. C., 1887, Notice of a new gigantic dinosaur: American Journal of Science, 3rd series, v. 24, p. 191-198.

Mateus, O., 2008, Two ornithischian dinosaurs renamed: *Microceratops* Bohlin 1953 and *Diceratops* Lull 1905: Journal of Paleontology, v. 82, n. 2, p. 423.

McDonald, A. T., and Horner, J. R., 2010, New material of "*Styracosaurus*" *ovatus* from the Two Medicine Formation of Montana: In: New Perspectives on Horned Dinosaurs. The Royal Tyrrell Museum Ceratopsian Symposium, edited by Ryan, M. J., Chinnery-Allgeier, B. J., and Eberth, D. A., Indiana University Press, Part Two, p. 156-168.

Ostrom, J. H., and Wellnhoffer, P., 1990, *Triceratops*: an example of flawed systematics: In: Dinosaur Systematics, Approaches and Perspectives, edited by Carpenter, K., and Currie, P. J., Cambridge university Press, p. 245-254.

Ott, C. J., and Larson, P. L., 2010, A new, small ceratopsian dinosaur from the Latest Cretaceous Hell Creek Formation, northwest South Dakota, United States, a preliminary description: In: New Perspectives on Horned Dinosaurs. The Royal Tyrrell Museum Ceratopsian Symposium, edited by Ryan, M. J., Chinnery-Allgeier, B. J., and Eberth, D. A., Indiana University Press, Part Two, p. 203-218.

Parks, W. A., 1925, *Arrhinoceratops brachyops*, a new genus and species of Ceratopsia from the Edmonton Formation of Alberta: University of Toronto Studies, Geological series, n. 19, 15pp.

Penkalski, P. G., 1994, The morphology of *Avaceratops lammersi*, a primitive ceratopsid from the Judith River Formation (Late Campanian) of Montana: Thesis, 55pp.

Penkalski, P. G., and Dodson, P., 1999, The morphology and systematics of *Avaceratops*, a primitive horned dinosaur from the Judith River Formation (Late Campanian) of Montana, with the description of a second skull: Journal of Vertebrate Paleontology, v. 19, n. 4, p. 692-711.

Rogers, K., 1991, The Sternberg fossil hunters A Dinosaur Dynasty: Mountain Press Publishing Company, 288pp.

Ryan, M. J., 2007, A new basal centrosaurine ceratopsid from the Oldman Formation, southeastern Alberta: Journal of Paleontology, v. 81, n. 2, p. 376-396.

Ryan, M. J., Eberth, D. A., Brinkman, D. B., Currie, P. J., and Tanke, D. H., 2010, A new *Pachyrhinosaurus*-like ceratopsid from the Upper Dinosaur Park Formation (Late Campanian) of southern Alberta, Canada: In: New Perspectives on Horned Dinosaurs. The Royal Tyrrell Museum Ceratopsian Symposium, edited by Ryan, M. J., Chinnery-Allgeier, B. J., and Eberth, D. A., Indiana University Press, Part Two, p. 141-155.

Ryan, M. J., Evans, D. C., Currie, P. J., and Loewen, M. A., 2014, A new chasmosaurine from northern Laramidia expands frill disparity in ceratopsid dinosaurs: Naturwissenschafte, published on line, 24, May, 2014, 8pp.

Ryan, M. J., Evans, D. C., and Shepherd, K. M., 2012, A new ceratopsid from the Foremost Formation (middle Campanian) of Alberta: Canadian Journal of Earth Sciences, v. 49, p. 1251-1262.

Ryan, M. J., Holmes, R., and Russell, A. P., 2007, A revision of the Late Campanian Centrosaurine Ceratopsid genus *Styracosaurus* from the Western Interior of North America: Journal of Vertebrate Paleontology, v. 27, n. 4, p. 944-962.

Ryan, M. J., and Russell, A. P., 2005, A new centrosaurine ceratopsid from the Oldman Formation of Alberta and its implications for centrosaurine taxonomy and systematics: Canadian Journal of Earth Sciences, v. 42, p. 1369-1387.

Sampson, S. D., 1995, Two new horned dinosaurs from the Upper Cretaceous Two Medicine Formation of Montana; with a phylogenetic analysis of the Centrosaurinae (Ornithischia: Ceratopsidae): Journal of Vertebrate Paleontology, v. 15, n. 4, p. 743-360.

Sampson, S. D., and Loewen, M. A., 2010, Unraveling a radiation: a review of the diversity, stratigraphic distribution, biogeography, and evolution of horned dinosaurs (Ornithischia: Ceratopsidae): In: New Perspectives on Horned Dinosaurs. The Royal Tyrrell Museum Ceratopsian Symposium, edited by Ryan, M. J., Chinnery-Allgeier, B. J., and Eberth, D. A., Indiana University Press, Part Four, p. 405-427.

Sampson, S. D., Loewen, M. A., Farke, A., Smith, J., and Roberts, E., 2009, Two new chasmosauine ceratopsids from Late Cretaceous (Campanian) of Utah: In: Journal of Vertebrate Paleontology, v. 29, supplement to n. 3, 69th Annual Meeting, Society of Vertebrate Paleontology and 57th Symposium of Vertebrate Paleontology and Comparative Anatomy (SPVCA): 175a.

Sampson, S. D., Loewen, M. A., Farke, A. A., Roberts, E. M., Forster, C. A., Smith, J. A., and Titus, A. L., 2010, New horned dinosaurs from Utah provide evidence for intracontinental dinosaur endemism: Public Library of Science (PLOS) One, v. 5, issue 9, 12 pp.

Schafer, W., 1976, Fossilien, objekte der erkenntnis, der praxis und der bildung: Nature und Museum, v. 106, n. 3, p. 65-73p.

Schlaikjer, E. M., 1935, Contributions to the Stratigraphy and Palaeontology of the Goshen Hole Area, Wyoming. No. II. The Torrington Member of the Lance Formation and a Study of a New *Triceratops*: Bulletin of the Museum of Comparative Zoology, Harvard, v. 76, p. 31-68.

Sternberg, C. M., 1929, A new species of Horned Dinosaur from the Upper Cretaceous of Alberta: Bulletin of the National Museum of Canada, n. 54, p. 34-37.

Sternberg, C. M., 1938, *Monoclonius* from southeastern Alberta compared with *Centrosaurus*: Journal of Palaeontology, v. 12, n. 3, p. 284-286.

Sternberg, C. M., 1940, Ceratopsidae from Alberta: Journal of Palaeontology, v. 14, n. 5, p. 468-480.

Sternberg, C. M., 1949, The Edmonton fauna and description of a new *Triceratops* from the Upper Edmonton Member: phylogeny of the ceratopsidae: Annual Report of the National Museum, Bulletin n. 113, p. 33-46.

Sternberg, C. M., 1950, *Pachyrhinosaurus canadensis*, representing a new family of the Ceratopsia, from Southern Alberta: Bulletin of the National Museum of Canada, v. 118, p. 109-114.

Sullivan, R. M., and Lucas, S. G., 2010, A new chamosaurine (Ceratopsidae, Dinosauria) from the Upper Cretaceous Ojo Alamo Formation (Naashoibito Member), San Juan Basin, New Mexico: In: New Perspectives on Horned Dinosaurs. The Royal Tyrrell Museum Ceratopsian Symposium, edited by Ryan, M. J., Chinnery-Allgeier, B. J., and Eberth, D. A., Indiana University Press, Part Two, p. 169-180.

Tokaryk, T. T., 1986, Ceratopsian Dinosaur from the Frenchman Formation (Upper Cretaceous) of Saskatchewan: Canadian Field-Naturalist, v. 100, n. 2, p. 192-196.

Tyson, H., 1977, Functional craniology of the Ceratopsia (Reptilia: Ornithischia) with special reference to *Eoceratops*: The University of Alberta, Facility of Graduate studies and research, Thesis, 339pp.

Ukrainsky, A. S., 2007, A new replacment name for *Diceratops* Lull, 1905 (Reptilia: Ornithischia: Ceratopsidae): Zoosystematica Rossica, v. 16, n. 2, p. 292.

Ukrainsky, A. S., 2009, Synonymy of the Genera *Nedoceratops* Ukrainsky, 2007 and *Diceratus* Mateus, 2008 (Reptilia: Ornithischia: Ceratopidae: Palaeontological Journal v. 43, n. 1, p. 116.

Wick, S. L., and Lehman, T. M., 2013, A new ceratopsian dinosaur from the Javelina Formation (Maastrichtian) of West Texas and implications for Chasmosaurine phylogeny: Naturwissehschaften, published on line, 16pp.

Wiman, C., 1930, Uber Ceratopsia aus der Obered kreide in New Mexico: Nova Acta Regiae Societatis Scientiarum Upsaliensis, 4th series, v. 7, n. 2, p. 3-19.

Wolfe, D. G., Kirkland, J. I., Smith, D., Poole, K., Chinnery-Allgeier, B., and McDonald, A., 2010, *Zuniceratops christopheri*: the North American ceratopsid sister taxon reconstructed on the basis of new data: In: New

Perspectives on Horned Dinosaurs. The Royal Tyrrell Museum Ceratopsian Symposium, edited by Ryan, M. J., Chinnery-Allgeier, B. J., and Eberth, D. A., Indiana University Press, Part Two, p. 91-98.

Wu, X.-C., Brinkman, D. B., Eberth, D. A., and Braman, D. R., 2007, A new ceratopsid dinosaur (Ornithischia) from the uppermost Horseshoe Canyon Formation (upper Maastrichtian), Alberta, Canada: Canadian Journal of Earth Sciences, v. 44, p. 1243-1265.

Xu, X., Wang, K., Zhao, X., and Li, D., 2010, First ceratopsid dinosaur from China and its biogeographical

For the Dinosaur Enthusiast and Collector

PRE HISTORIC TIMES

August/September
2001 Issue NO. 49

Interviews & Articles
with Dinosaur Sculptors
& Modelers

Jurassic Park III

When Dinosaurs
Roamed America

How to Build Your
Own Dinosaur Models

U.S. $5.95 • Canada $6.95

28

Ford, T. L., 2001, How to Draw Dinosaurs. 6 pack abs (times 5), or wash board abs: Prehistoric Times, v. 49, p. 14 - 15.

Chapter 3

6 pack abs (times 5), or wash board abs

The topic of this issue was inspired Phil Currie. I was sitting next to Mike Skreptnick during his Sunday talk about Feathered Dinosaurs (held at the Chicago Field Museum, May 13, 2001). Phil showed two pictures of completely articulated ornithmimids including their gastralia (or belly ribs). They were in a natural position, which intrigued both he and I. Mike leaned over and said "look, the gastralia are in a straight line, not bowed", and with that, this article was born.

I had thought about doing an article on this before, but after that talk and the talk by Leon Classens (The function of gastralia in theropod lung ventilation at the A. Watson Armour III Symposium, The Paleobiology and Phylogenetic of Large Theropods, also held at the Field Museum) I thought, yep, that's the next article. Gastralia are also called belly ribs or the gastric basket, because of their location: the belly. Gastralia are not unique to theropods, they go back to the earliest tetrapods. The only living tetrapods with gastralia are the crocodilia and the tuatara. Theropods, prosauropods, and early birds (and possibly some sauropods, but that is in debate) had gastralia. Mammals lack gastralia (I'd imagine that if we did have belly ribs you'd be seeing inframerisals on getting real wash board abs!).

I'm don't know what the gastralia function was in the early tetrapods, but in theropods (at least) are believed to have to do with breathing. I will be using *Gorgosaurus libratus* in the illustrations because this is one of the best theropod gastralia scientifically illustrated (Figure 1). The gastralia were incased in muscles (just inside the skin and completely covered in musculature) and below the 'guts'. They start just behind the sternum and in many cases go all the way down to the pubis. The gastralia are paired (offset) long rod-like structures with the 'pointed' end going away from the belly (or toward the lower edge of the true ribs) with the first (one or more, some anterior gastralia had flatter plate like look) and last a solid boomerang (Figure 2). The paired ribs interlock and have a 'sliding' joint.

According to Leon, the ribs slid against each other during breathing. They would slide out during inhalation, and in for exhalation, but not to a great degree (Figure 3). They didn't bow upward from the pubis as some have depicted (Figure 4a). The artists can show the belly in a variety of degrees of breathing. With more than one animal, one could have the gastralia slightly concave (showing inhaling Figure 4b) and others convex (showing exhaling Figure 4c)

Finally, like true ribs, gastralia can be broken in life and healed. One such case had confused Charles Gilmore into believing *Allosaurus* had several different gastralia instead of a single rod (Dan Chure recently wrote on this). Also, some furculae have been misidentified as gastralia (I'll save this for a possible later article).

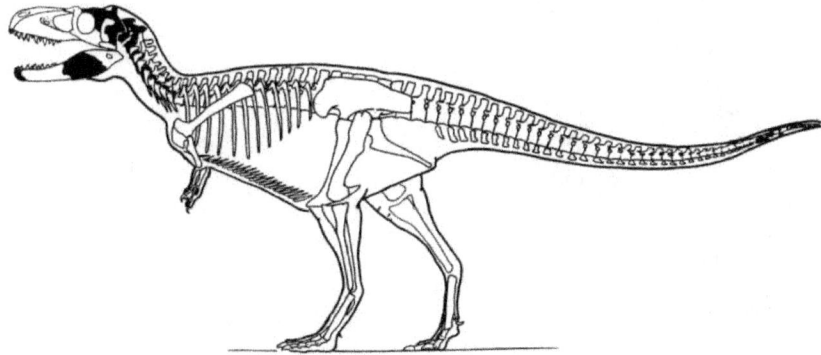

Figure 1): Skeletal restoration of *Gorgosaurus libratus*

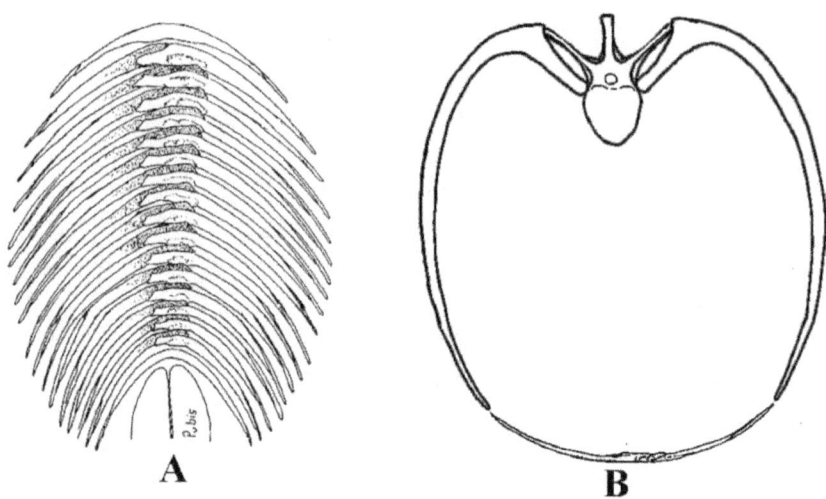

Figure
2): Gastralia of *Gorgosaurus libratus;* a) in place (all after Lambe, 1917); b) cross section showing the ribs and gastralia (after Lambe, 1917).

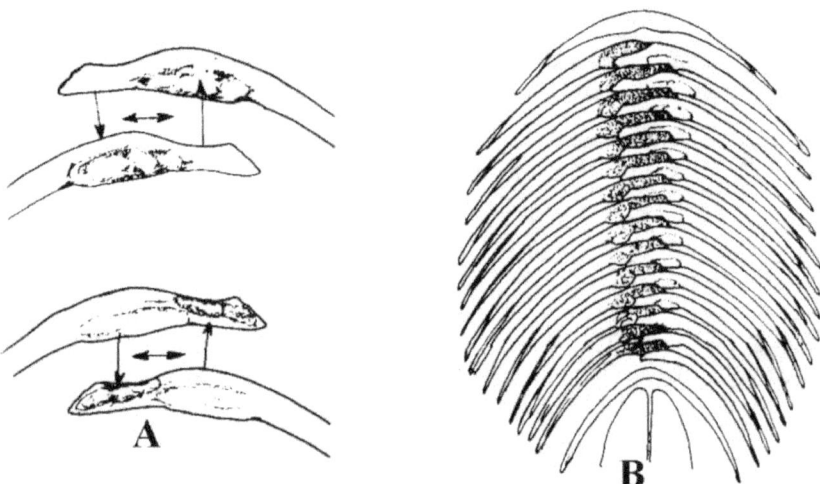

Figure
3) Gastralia showing; a) the joint; b); and the gastralia expanded (modified from Lambe, 1917).

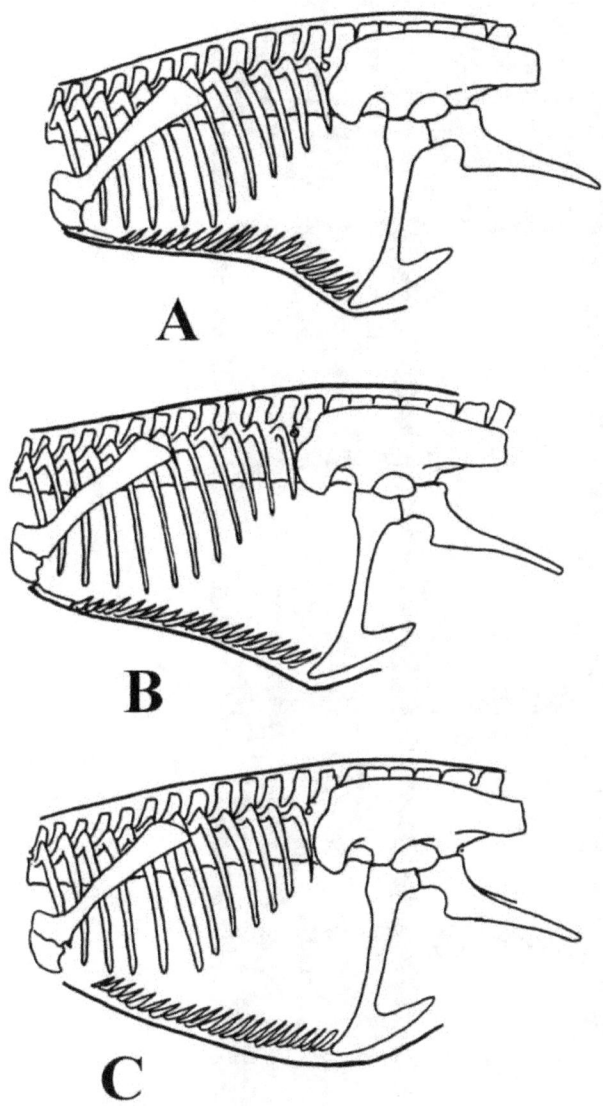

Figure 4) Gastralia shown in different view; a) incorrect extension; b) inhalation; c) exhalation.

Bibliography

Lambe, L. M., 1917, The Cretaceous theropodous dinosaur *Gorgosaurus*: Geological Survey of Canada Memoir, v. 100, p. 1-84.

For the Dinosaur Collector and Enthuiast

PREHISTORIC TIMES

Oct/Nov 2001

BIG 50TH COLLECTOR'S ISSUE

US $5.95 • Canada $6.95

This Issue:
- Tyrannosaurus rex
- Woolly Rhinos
- Alan Dean Foster
- Philip J Currie
- Mark Hallett
- Larry Martin
- John R Horner

32

Ford, T. L., 2001, How to Draw Dinosaurs. Sails in the Mesozoic: **Prehistoric Times, n. 50, p. 14-15.**

Chapter 4

Sails in the Mesozoic

(First off, I'd like to let the readers know that my second volume of How to Draw Dinosaurs is out for $20, plus postage. Shameless plug).

With the opening of Jurassic Park 3, I thought it would be appropriate to dedicate this issues to dinosaurs with sails. Not all dinosaurs known to have had a sail had neural arches (which make up the sail) of the vertebrae went straight up. They were not rod-like (as in pelycosaurs) but were flat and blade-like. *Spinosaurus*, the scourge of JP3 is shown to have had a sail that had straight neural spines and a rounded sail. This was, no doubt partly attributed to Paul Sereno (et. al.) paper on *Suchomimus* in which they depicted *Spinosaurus* with such a sail. This is completely incorrect. The front dorsal spines are tilted forward and are short with the spines growing in height nearer to the pelvis. The base of the spines has a small 'bulge' (which might have been covered with muscle to that area). Some spines have a 'wave' to them (when viewed front to back, see figure 1a). It is possible that Spino's curvy dorsal spines are due to movement of the earth over the millions of years and didn't actually look that way in life? I've talked to some people about that and they think no. It wasn't deformed during fossilization. Unfortunately only a caudal vertebra is known and its exact position isn't known (this caudal is very important, in that it can determine how the sail was). If the caudal was nearer to the pelvis then the sail was short (figure 2b). If it was near the middle of the tail then the sail would have been tapered more or less like *Suchomimus* (figure 2a). If it were further back then the sail would have been very long (similar to *Ouranosaurus*, 2c). Another important thing to look at is the ribs. The ribs of *Spinosaurus* indicate that the animal was in fact, a very wide animal (figure 1b).

Suchomimus had a short sail with the tallest part of the sail over the pelvis (figure 3). If the 'sail' was covered with muscle (as some have theorized) then the tail would have been very heavy and thick. I doubt the animal's sail was completely covered with muscle.

One of disappointing aspects of paleontology is the incompleteness of the fossil record as well as the incompleteness of the skeleton itself. The new *Acrocanthosaurus* skeleton is one such case. NCSM 14345 was found in McCurtin County, Oklahoma (figure 4a). It has the best skull known for *Acrocanthosaurus*. *Acrocanthosaurus* had a short sail, in this case being so short it was probably enclosed in muscle. The skeleton (NCSM 14345) itself lacks complete dorsal vertebrae (with one or two dorsal neural spines). Therefore the exact height of the back is unknown. Another less complete specimen known from a partial skeleton, SMU 74646, was found in Parker County (figure 4c), Texas. This specimen has a bit more material than the type (MUO 8-0-S9, also found in Oklahoma, figure 4c). The new mounted skeleton is cool to look at but I'm not sure I agree with the arrangement of the dorsal neural spines.

Theropods aren't the only dinosaurs with a sail; *Ouranosaurus* (also found in Northern Africa as was *Spinosaurus*) had a tall sail (not as tall as *Spinosaurus*), which tapered toward the tail (figure 5a). *Vectisaurus*, from the Early Cretaceous of England (referred material illustrated, *Vectisaurus* is considered by many to be juvenile *Iguanodon*) had a shorter sail (figure 5b). ?*Lambeosaurus laticaudus* (editor's note, now called *Magnapaulia laticaudus* Prieto-Marquez, Chiappe, & Joshi, 2012), from Baja California, had the tallest 'sail' of any known hadrosaur (figure5c). In some hadrosaurs the tallest part of the 'sail' was on the tail; as with *Shantungosaurus* (figure 5d)

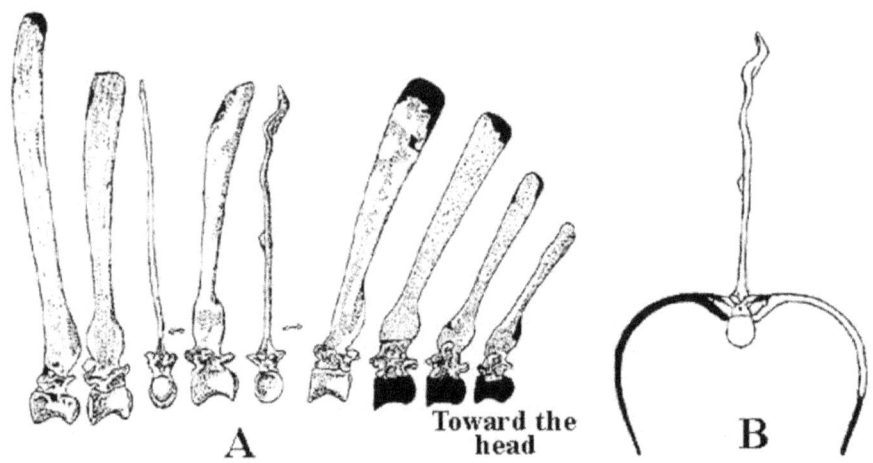

Figure

1) *Spinosaurus*; A) dorsal vertebrae; B) width of *Spinosaurus* (after Stromer, 1915).

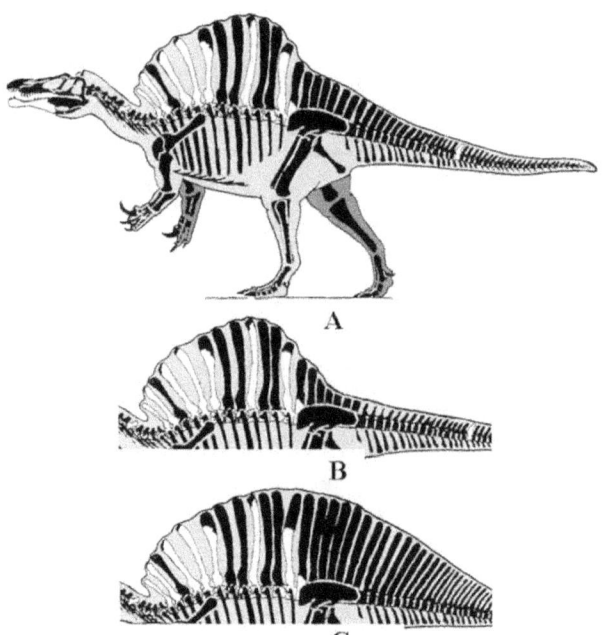

Figure 2A) Skeleton of *Spinosaurus* with a medium sail like *Suchomimus*; B) short sail; C) long sail like *Ouranosaurus*.

34

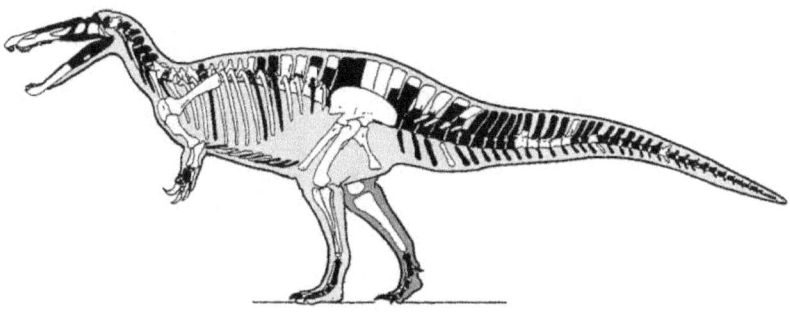

Figure 3) Skeleton of *Suchomimus* (modified from Sereno, et al., 1998) .

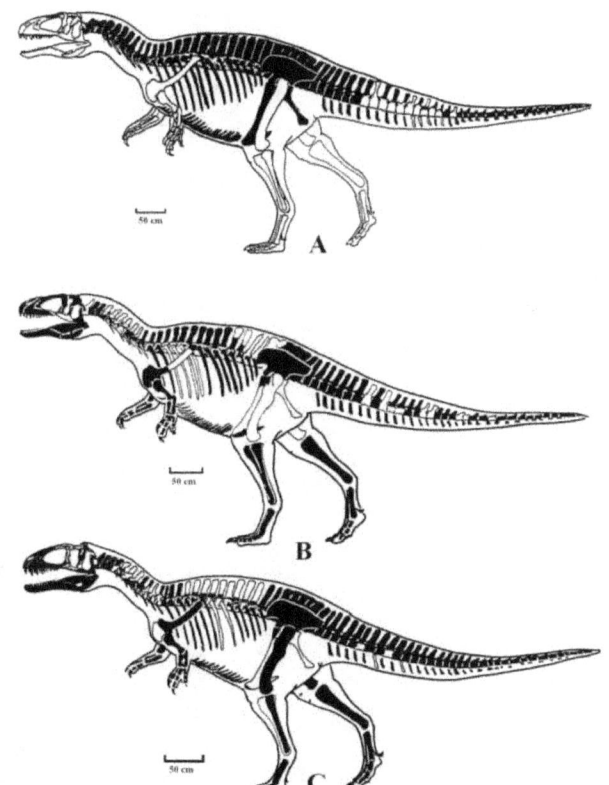

Figure 4) *Acrocanthosaurus* A) NCSM 14345 (modified from Currie & Carpenter, 2000); B) SMU 74646 (modified from Harris,1997, 1998); C) MUO 8-0-S9 (modified from Stovall & Langston, 1950).

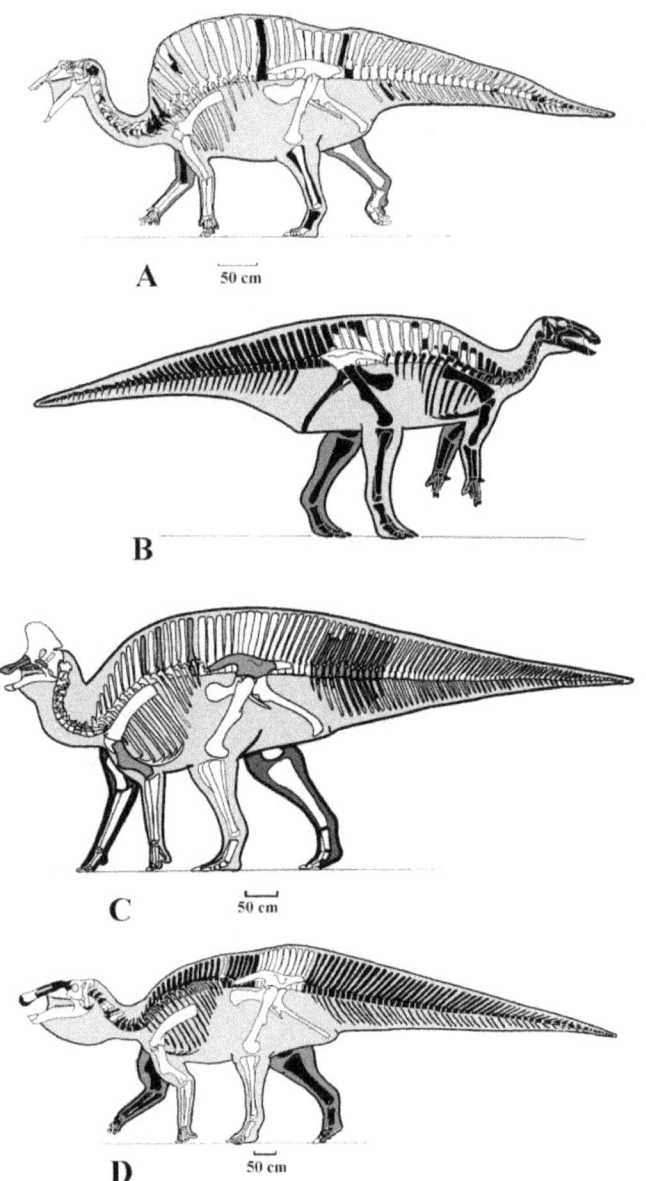

Figure 5A) *Ouranosaurus* (modified from Taquet, 1976); B) *Vectisaurus* (modified from Galton, 1976); C) ?*Lambeosaurus laticaudus* (modified from Prieto-Marquez, et al., 2014); D) *Shantungosaurus* (modified from Hu, 1973, Hu, et al., 1988, Hu, et al., 2001).

Bibliography

Currie, P. J., and Carpenter, K., 2000, A new specimen of *Acrocanthosaurus atokensis* (Theropoda, Dinosauria) from the Lower Cretaceous Antlers Formation (Lower Cretaceous, Aptian) of Oklahoma, USA: Geodiversitas, v. 22, n. 2, p. 207-246.

Galton, P. M., 1976, The dinosaur *Vectisaurus valdensis* (Ornithischia: Iguanodontidae) from the Lower Cretaceous of England: Journal of Paleontology, v. 50, n. 5, p. 976-984.

Harris, J. D., 1997, A reanalysis of *Acrocanthosaurus atokensis*, its phylogenetic status, and paleobiogeographic implications, based on a new specimen from Texas: A thesis Presented to the Graduate Faculty of Dedman College Southern Methodist University in Partial Fulfillment of the Requirements for the degree of Master of Science with a Major in Geological Sciences, p. 1-204.

Harris, J. D., 1998, A Reanalysis of *Acrocanthosaurus atokensis*, its phylogenetic status, and paleobiogeographic implications, based on a new specimen from Texas: New Mexico Museum of Natural history and Sciences, Bulletin n. 13, p. 1-75.

Hu, C. C., 1973, A new Hadrosaur from the Cretaceous of Chucheng, Shantung: Acta Geologica Sinica, n. 2, p. 179-206.

Hu, C. C., and Cheng, Z., 1988, New Progress in the Restudy on *Shantungosaurus giganteus*: Bulletin of CAGS, p. 251-258.

Hu, C. C., Cheng, Z., Pang, Q., and Fang, X., 2001, *Shantungosaurus giganteus*: Geological Publishing House, 139pp.

Morris, W. J., 1981, A new species of hadrosaurian dinosaur from the Upper Cretaceous of Baja California- ?*Lambeosaurus laticaudus*: Journal of Paleontology, v. 55, n. 2, p. 453-462.

Prieto-Marquez, A., Chiappe, L. M., and Joshi, S. H., 2012, The lambeosaurine dinosaur *Magnapaulia laticaudus* from the Late Cretaceous of Baja California, northwestern Mexico: Public Library of Science (PLOS), One, v. 7, n. 6, 32pp.

Sereno, P. C., Beck, A. L., Dutheil, D. B., Gado, B., Larsson, H. C. E., Lyon, G. H., Marcot, J. D., Rauhut, O. W. M., Sadleir, R. W., Sidor, C. A., Varricchio, D. D., Willson, G. P., and Wilson, J. A., 1998, A Long-Snouted Predatory Dinosaur from Africa and the Evolution of Spinosaurids: Science, v. 282, p. 1298-1302.

Stovall, J. W., and Langston, W. Jr., 1950, *Acrocanthosaurus atokensis*, a New Genus and Species of Lower Cretaceous Theropods From Oklahoma: American Midland Naturlist, v. 43, n. 3, p. 696-728.

Stromer, E., 1915, Ergebnisse der Forschungsreisen Prof. E. Stromers in den Wusten Agypten s. II. Wirbeltier-Reste der Baharije-Stufe (unterstes Cenoman). 3. Das Original des Theropoden *Spinosaurus aegyptiacus*: Abhandlungen der Koniglich Bayerischen Akademie der Wissenschaften Mathematisch-physikalische Klasse 28, band 3, p. 3-32.

Taquet, P., 1976, Geologie et Paleontologie du gisement de Gadoufaoua (Aptian du Niger): Cahires Paleont, 191pp.

I would like to expand How to Draw Dinosaurs to cover non dinosaurs. I'll call this column "How to Draw Non-Dinosaurs", for now this will appear occasionally.

This first article is on my favorite prehistoric animals (after dinosaurs of course) pelycosaurs (in keeping with sails). I had planed on writing on this sooner but, luckily, the article didn't get published. I was, at that time, under the impression that pelycosaurs lacked the fleshy sail, but had a rod-like 'sail'. I found a paper by a German paleontologist (Jaekel) who had the same idea, but I soon dropped that idea.

In April when I was at the Chicago Field Museum for their Theropod symposium, I had the pleasure of talking to Stuart Sumida who was in the collections at the time. We talked about pelycosaurs and how they had to have had a 'sail' with skin. He was looking for pathologies (i.e. broken sails) of pelycosaurs (for research that his wife was doing, and I did find a couple of pathologies for them in the collections). If there were no skin in the sail, then the sail would have fallen off a not been able to heal itself. So, there was a sail.

Recently Adrian Hunt, and Spencer Lucas (1998) have come to the conclusion that pelycosaurs had a more erect gait and not the more typical sprawling one. There are two main bodies of evidence for this; first is the footprint evidence, the other the olecranon. The majority of pelycosaurs have a large olecranon (or extension on the ulna, figure 1). Many quadrapedal mammals with an erect gait have this large olecranon (including stegosaurs and ceratopians). The large attachment area is for muscles that help the animal in maintain an erect gait. The footprints are closely placed to one another (*Dimetropodus*, figure 2 c) and not far apart. This indicates that the animal had its feet closer to the midline and had to have had an erect gait.

Another possibility for an erect gait is the sail itself. If you take two dowels and place them in rubber, then try to separate the two while holding onto the bottom of the dowels you'll have a difficult time (granted, skin would have been easier). If you do it with a dozen or so dowels it becomes increasingly harder due to this great support. This may be what is going on in pelycosaurs. So, the old view of a sprawling pelycosaur is incorrect and a new more erect pelycosaur is the norm.

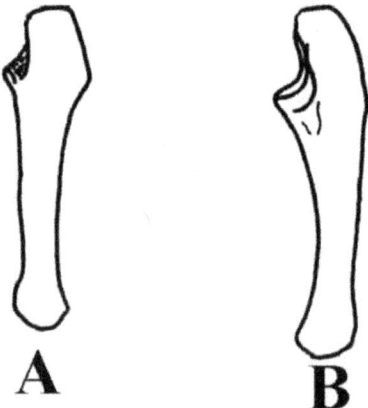

Figure 1A) Ulna of *Dimetrodon grandis*: B) *Edaphosaurus boangeres* (after Romer, & Price, 1940)

Figure
2A) Skeleton of *Dimetrodon grandis* and *Edaphosaurus crucigeri* in a sprawling pose (modified from Romer & Price, 1940); B) both in a more erect pose (modified from Romer & Price, 1940); C) trackway of *Dimetropodus* (after Hunt & Lucas, 1998).

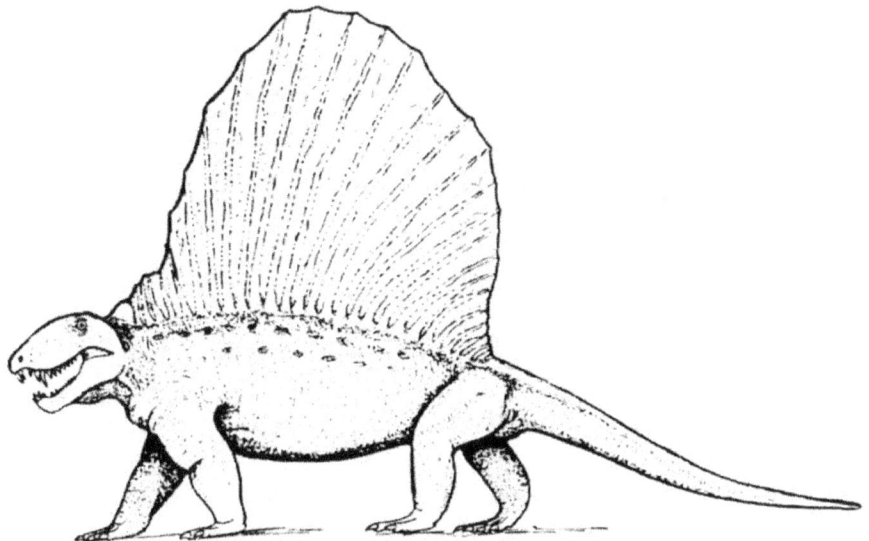

Figure 3) *Dimetrodon* is now believed to have walked with a more erect posture than a sprawling one.

Bibliography

Hunt, A. P., and Lucas, S. G., 1998, Vertebrate tracks and the myth of the belly-dragging, tail-dragging tetrapods of the Late Paleozoic: In: Permian Stratigraphy and Paleontology of the Robledo Mountains, New Mexico, edited by Lucas, S. G., Estep, J. W., and Hoffer, J. M., New Mexico Museum of Natural History and Sciences, Bulletin n. 12, p. 67-69.

Romer, A. S., and Price L. I., 1940, Review of the Pelycosauria: Geological Society of America Special Papers n. 28, p. 1-538.

For the Dinosaur Collector and Entusiast

PREHISTORIC TIMES

#51 Dec/Jan 2002

Building A
LIFESIZE
T. REX

Dinosaur
Photographer
LOUIE
PSIHOYOS

US $5.95 • Canada $6.95

Dinosaur Reviews,
Art & Articles

41

Ford, T. L., 2002, How to Draw Dinosaurs. To know the nose, part 1: Prehistoric Times, n. 51, p. 14-15.

Chapter 5

To know the nose, part 1

This article is a correction to an article I wrote for Prehistoric Times a few years ago (Number 40: 14-15, The Theropod nose). As I've done in the past I will make any corrections that are needed to help everyone keep informed and updated on new theories. Some of what I will write will be similar to what was written in the previous article, but it is important to understand what is being theorized.

Larry Witmer recently published a new theory on dinosaur noses (in the journal Science) and gave a talk at this year's SVP (see bibliography). He and his collogues have been studying modern animal noses; reptiles, birds, and mammals and have come to some preliminary conclusions. I will be relaying those conclusions and await further work on their behalf.

This will be a 3-part article; part one will be on theropods and prosauropods; part 2 will cover the ornithischians and part 3 will be on sauropods. Part three will come out when Witmer's (and colleagues?) paper on sauropod noses is released and I have a better understanding of his/their conclusions.

I originally stated that I believed that the nose of theropods would have been similar to birds; with a large 'external hole' where the naris is (figure 1). I did not know that the external naris is not the exact size of the skeletal naris. I explained that the majority of theropods have a 'bar' (consisting of the premaxilla and anterior portion of the nasals) at the front of the nose which would have had a narrow area for the external narial opening.

Witmer has x-rayed, scanned, and dissected several animal heads and have come to the conclusion that the 'primitive' condition for the fleshy external nose is to have the narial opening low and toward the front of the snout (figure 2). When I talked to him about his theory he stated that the narial opening itself is misleading and helps to confuse people to the actual external opening of the nose (and I will explain why in all three parts).

The largest part of the narial opening is at the posterior end. It fills the upper posterior part of the naris. Think of it like this, look at a skull of a mammal and you'll see from side view you can't see the narial opening, but looking at head on you'll see the narial opening (with lots of cartilage). Dinosaurs are basically the same, but the majority of them have a 'narial bar'. This is where the nasal passage passes through to the inside of the skull. The skeletal portion of the naris is made up of the premaxilla and nasal bones (both a left and right, and sometimes a part of the maxilla).

In theropods the narial opening is narrow and forms a 'bar' (as I stated before). The ventral edge of the narial opening has a shelf (this shelf can vary from thin to rather wide, depending on what theropod/prosauropod) (figure 3). I was at a loss for the purpose of this shelf. Thanks to Witmer I am no longer at a loss.

What Witmer is saying is that the external narial opening itself was placed at the ventral front end of the nasal area (skeletal area or at the lower front end of the bar) and that soft tissue covered the nose from the bar to the shelf (figure 4). The purpose of the soft tissue isn't known (right now). Think of it like this; the bar in theropods/prosauropods would have been similar to the cartilage in our nose (and mammals) and the soft tissue would have been from the bar to the shelf making the nose round and blocky (the opposite of what I said before). If you broke off the 'bar' you'd have a theropod/prosauropod that would look like a mammals (cover the 'bar' of any skull of a theropod). So the large narial opening in theropods/prosauropods wouldn't have been as large when it was covered with soft tissue and the naris would have been lower and closer to the tip of the snout. The external naris was not regulated by the skeletal naris opening, but would have extended lower onto the 'shelf' area. Whether or not there was any muscle tissue that could close the nose off is just speculation for now (figure 5). Also the external naris may have been larger than I've illustrated them.

Spinosaurids and dilophosaurids would have had a different nasal area because their narial opening is not at the front of the snout but further back (figure 5). Their nose would have been smaller and more 'flush' with the snout. The narial opening would have been placed closer to the front of the naris.

42

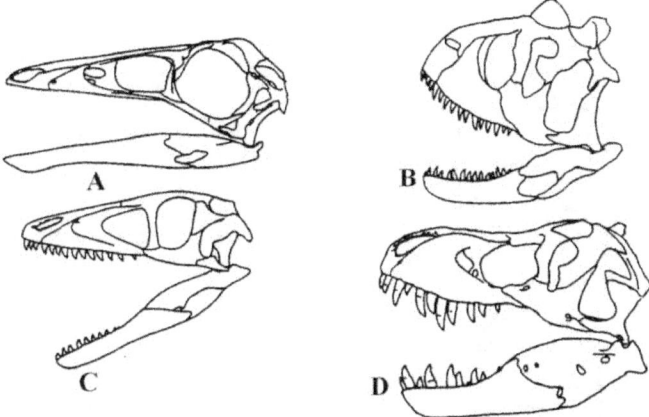

Figure 1). Skulls of four theropods; A) *Dromiceiomimus*; B) *Carnotaurus*; C) *Deinonychus*; D) *Tyrannosaurus*.

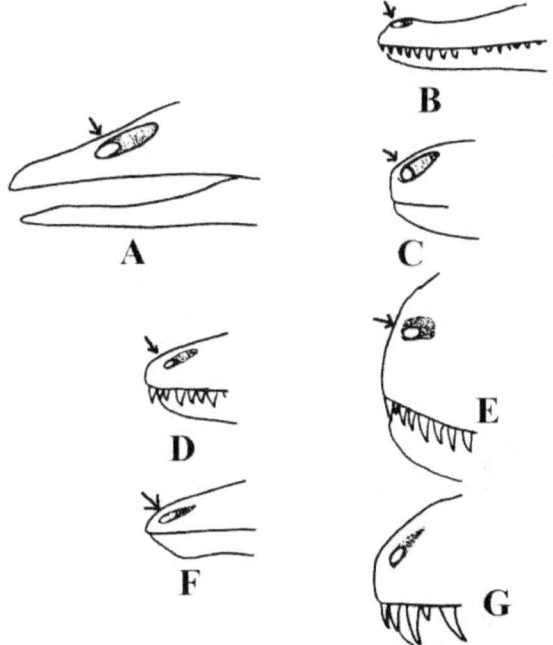

Figure 2). Outline of x-rays that were in Witmer's paper; A) a domestic goose; B) an American Alligator; C) spiny-tailed lizard and the re-interoperation of the nasal area of D) *Dromiceiomimus*; E) *Carnotaurus*; F) *Deinonychus*; G) *Tyrannosaurus*. The stippled area is where the narial opening of the skull is and the actual area of the exterior nasal opening.

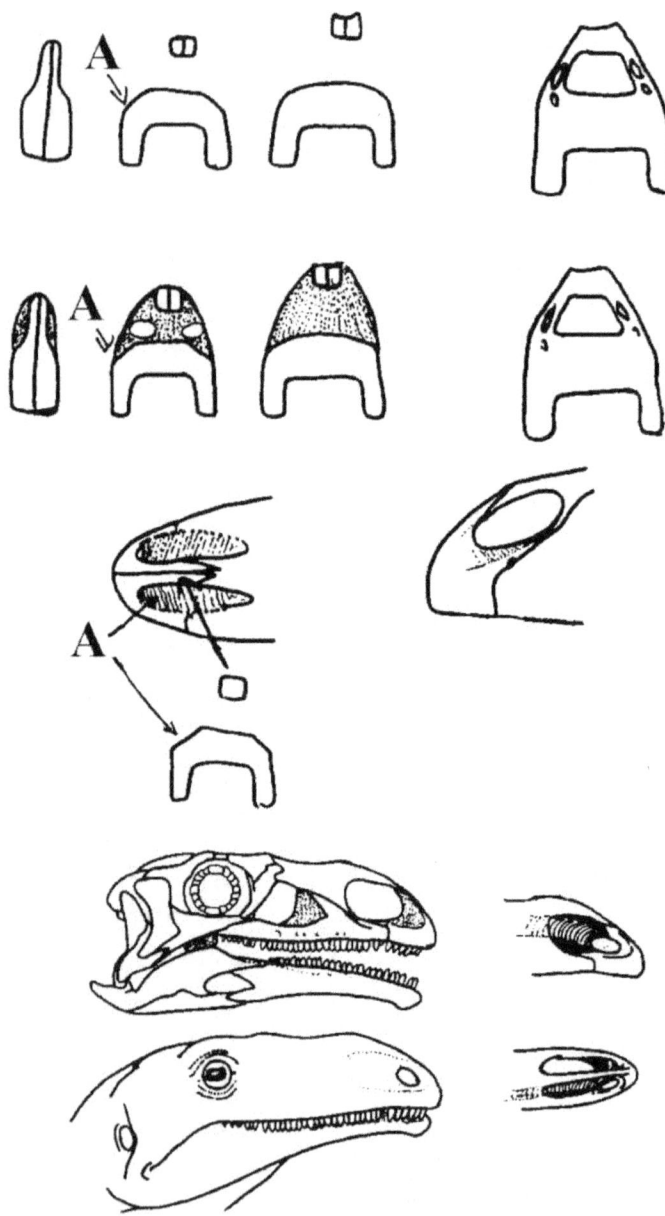

Figure
3). The narsal area showing the air passage to the inside of the skull and the 'shelf' (a) area of *Allosaurus* and showing the nasal area of *Plateosaurus* and interpretation of the nasal passage from the external nasal to the inside of the skull.

44

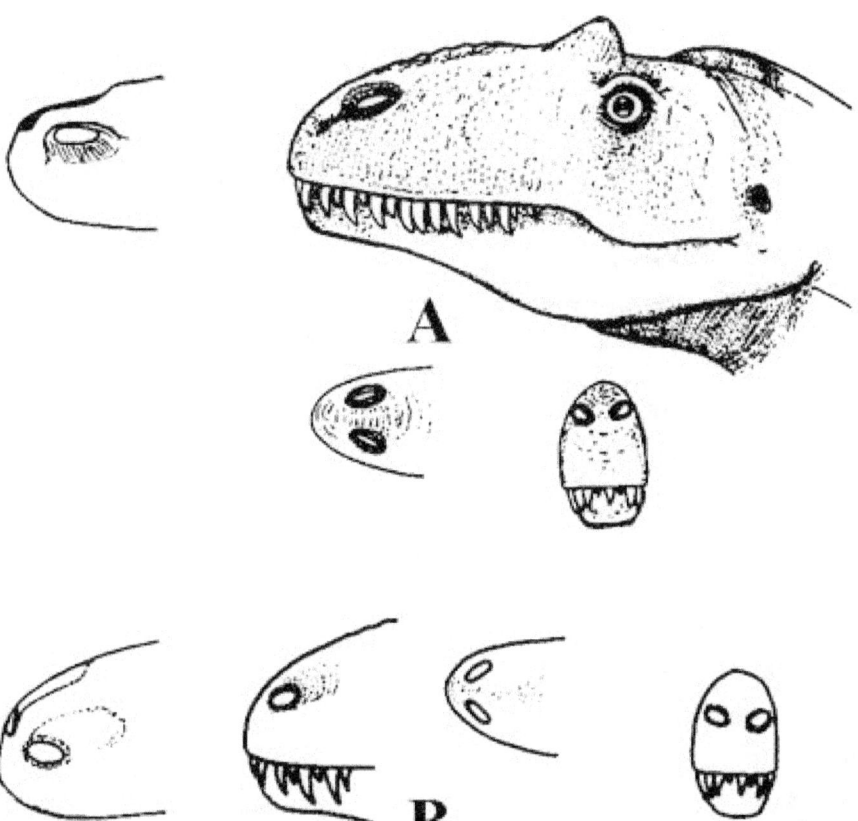

Figure 4). The nasal area showing A) how I first interpreted it (for *Allosaurus*) and B) showing Witmer's new interpretation and the soft tissue area covering the nasal area from the 'bar' to the 'shelf'.

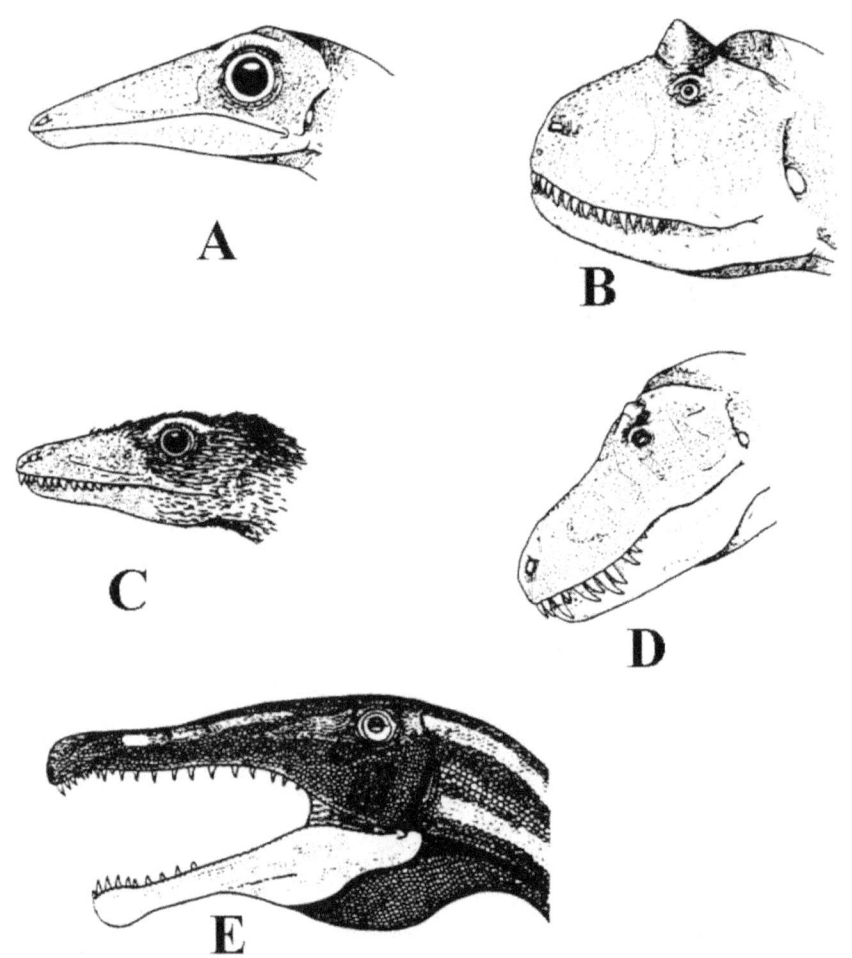

Figure
5). New interpretation of the nose of A) *Dromiceiomimus*; B) *Carnotaurus*; C) *Deinonychus*; D) *Tyrannosaurus*; E) the spinosaurid *Irritator*.

Bibliography

Witmer, L. M., 2001, Nostril position in dinosaurs and other vertebrates and its significance for nasal function: Science, v. 293, p. 850-853.

Witmer, L. M., 2001, The position of the fleshy nostril in dinosaurs and other vertebrates and its significance for nasal function: The Journal of Vertebrate Paleontology, Abstracts for papers, 61 annual meeting, Society of Vertebrate Paleontology, v. 21, supplement to n. 3, p. 115.

For the Dinosaur Collector and Enthusiast

PREHISTORIC TIMES

#52 Feb/Mar 2002

PaleoNews 2001
The Top Dino Finds of The Year

Sinclair Collectibles of The 1930's

Marcel Delgado's Dinosaurs

Dino Movies

E.H. Colbert's Last Interview

US $5.95 • Canada $6.95

0 56698 949 0

Ford, T. L., 2002, How to Draw Dinosaurs. To know the nose, part 2: Prehistoric Times, n. 52, p. 14-15.

Chapter 6

To know the nose, part 2

Continuing with the previous issue's article on the new theory about the nose of dinosaurs I will now cover the nose of ornithopods according to the recent research by Dr. Larry Witmer. According to Witmer, theropod's and prosauropod's noses don't look too different than before, but ornithopods are a totally different matter. After talking a bit with Larry I am better able to explain his views.

The large oval 'hole' or skeletal naris in ornithopods is deceptive (as per Witmer); in that it makes one believe that the external naris was somewhere inside that area (figure 1a); whether in front, top or back of that area. According to Witmer, the external naris would have been far forward from the skeletal naris. There is a 'groove' from the skeletal naris to the margin of the premaxilla in ornithopods. The premaxilla had a keratin covering (and will be covered in the next issue), as did the predentary that limited how far forward the external naris could go. There would have been some soft tissue, at least a few centimeters from the keratin border to the outside border of the fleshy external naris. Because of this border, hadrosaurids would have looked the most different with its 'fleshy' nose just behind the tip of the 'bill' (figure 1b-c). I doubt some hadrosaurids would have had the 'fleshy' bellowing noise like an elephant seal as has been depicted by others. This 'groove' has intrigued me in the past. I'd wonder what it was for. The area it is in is slightly indented and had to have been filled with something. It was most probably filled with soft tissue in the same way as with theropods. In primitive ornithopods the fleshy nasal-opening wasn't that far from the skeletal narial opening (figure 2a, b)

Lambeosaurids lack the large oval external naris, but the 'fleshy' nose would be near the tip of its head as its hadrosaurian cousins (figure 2 c).

Ceratopsians 'fleshy' nose would also have been further forward on the snout than is typically depicted and would be harder to accept because of the bizarre appearance it gives the animal (figure 2d, e). However, just because something looks strange doesn't mean it should be ignored.

I look forward to more of Dr. Witmer's work and how it may change the look of dinosaurs.

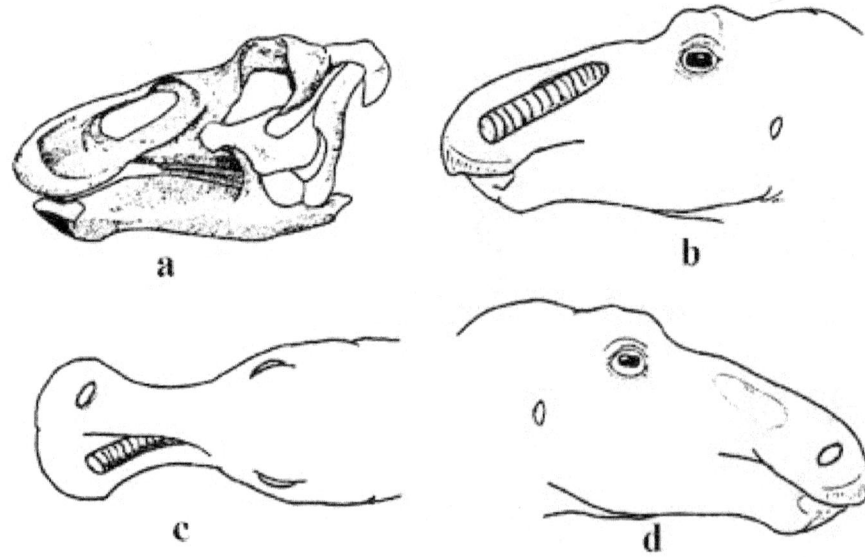

Figure 1a) Skull of *Edmontosaurus* showing the large skeletal nasal opening; b) side view of a fleshed out head showing where the 'nasal' tube would go; c) top view showing the 'nasal' tube and a fleshed out head; d) a fleshed out head showing the new position of the nasal.

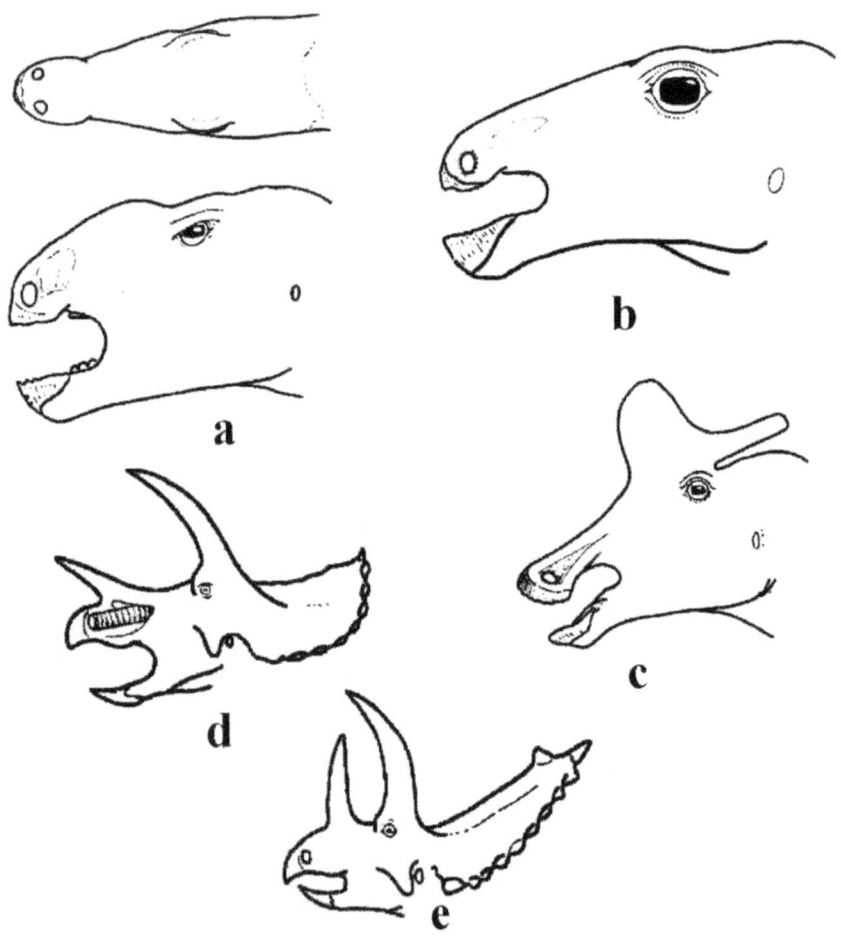

Figure
2) Fleshed out heads of ornithischian dinosaurs showing the position of the skeletal nasal opening shaded; a) *Tenontosaurus*, top and side view; b) *Iguanodon*; c) *Lambeosaurus*; d) head of a *Triceratops* showing the 'nasal' tube; e) a *Pentaceratops* head.

50

PREHISTORIC TIMES

#53 Apr/May 2002

Interview with
**Charles R
Knight**

Feathered
Dinosaurs

**Marx
Dinosaur
Playsets**

Elasmosaurus

Artist *Luis Rey*
and much more!

U.S. $5.95 • Canada $6.95

04

5669894980

Ford, T. L., 2002, How to Draw Dinosaurs. Duckbills or duckshovels?: Prehistoric Times, n. 53, p. 14.

Chapter 7

Duckbills or duckshovels?

As I mentioned in last time, this issue's article is on the rhamphotheca or the 'bill' of the duck-billed dinosaurs (or hadrosaurs). The reason why hadrosaurs have the nickname 'Duckbilled dinosaurs' is because of the large 'bill' that reminds people of a duck's bill. This bill varies in width and can be large or small (as I reported in PT #38). Only a few specimens of hadrosaurs have the rhamphotheca preserved. One is at the Los Angles Natural History Museum (LACM 23502, Figure 1) another at the Senckenburg Museum (Germany) (Figure 2, which is also mummified and I'll be writing about the 'mitt' next issue), and one that isn't really well known, at the Natural History Museum at the University of Michigan (at least these are the ones that I know of). I was able to photograph the LACM specimen 'through glass' but was pleasantly surprised that while visiting the Tucson Rock/Fossil/Mineral show, a cast of the Senckenburg specimen was in the lobby of the Ramada Inn. The cast is being sold by Glen Rockers of PaleoSearch and he was more than happy to let me photograph and study it.

The vast majority of skulls of hadrosaurs, both hadrosaurinae and Lambosaurinae, lack the rhamphotheca and the bill is usually illustrated without one.

First, just exactly what is a rhamphotheca? A rhamphotheca is 'horny' part of a bill. The hadrosaurs that have the rhamphotheca preserved show that it extends much further the fossilized bone and forms large flat surface. The inside has serrations or more correctly, ridges. The predentary has a serrated edge and may be part of the reason for the serrations on the inside of the rhamphotheca.

While writing an article for the Mesa Southwest Museum's Symposium (held late January, 2002) on the possibility of some dinosaurs being partly aquatic, I noticed some interesting things about jaws of hadrosaurs. It seems that lambeosaurs didn't have as large a rhamphotheca (Figure 1 e) as hadrosaurines (this may be partly due to preservation). Rhamphotheca has been reported from lambeosaurs but had been believed unimportant and destroyed when collecting the skull. The predentary of hadrosaurs is square (figure 3a, b) in dorsal/ventral view. There is a 'ridge' on the outside of the predentary and may have acted like a 'bumper' for the upper jaw. It is also where the lower margin of the rhamphotheca/upper jaw would have been when the jaw was closed. But there is a problem. I was talking to Neal Larson (of the Black Hills Institute) about the juvenile *Edmontosaurus* skull's premaxilla on the cast they were selling. The premaxilla lacks a rhamphotheca and has a serrated edge. He said they have an adult with a similar premaxilla. It may be possible that the large rhamphotheca is pathologic, similar to when a rodent has a misaligned front tooth and doesn't wear off and grows into a circle. If the rhamphotheca wasn't being worn off it'd keep growing, but all three specimens that have a rhamphotheca are the same length. It may also be possible that different genera have different rhamphotheca lengths.

The spring of 2001 I took a 3-week trip around the eastern half of the United States and Canada. One stop was at the Natures Museum of Canada (Ontario). As I walked around the dinosaur gallery I was looking at the mounted skeleton of *Hypacrosaurus altispinus*, and was fascinated by the front end of the open mouth. The upper jaw was slightly misplaced so there's a gap between the maxilla and premaxilla, but there was enough space between the two to see the predentary. It is from this that I started to think that the lower jaw looks more like a gomphotheres elephant than a duck's bill. The lower jaw in gomphotheres vary from straight to strongly curved, very similar to hadrosaurs. *Edmontosaurus, Anatotitan* (Figure 4b), *Prosaurolophus maximus* have very long straight lower jaws (similar to *Trilophodon (Amebelodon) fricki* Figure 4a). *Tsintaosaurus* (Figure 4g), and the majority of lambeosaurs have curved lower jaws (similar to *Trilophodon giganteus* (Figure 4c), *T. lulli* (Figure 4d), and *Rhynchotherium edense* (Figure 4e), *R. edense* (Figure 4f). The predentary in hadrosaurs have wide, squared off front edge (similar to gomphotheres) (Figure 3a, b). I believe that hadrosaurines and lambeosaurs ate differently. Hadrosaurines with the large rhamphotheca possibly used the upper jaw as a scrapper, dragging in large amounts of food, while lambeosaurines may have been shovelers, taking in large quantities of foot (Mike Brett-Surman believes lambeosaurines nipped their food). Their lower jaw is thicker than the front of the upper jaw and reminds me of a large shovel tractor with a large bucket and a small 'lid' that closes over it.

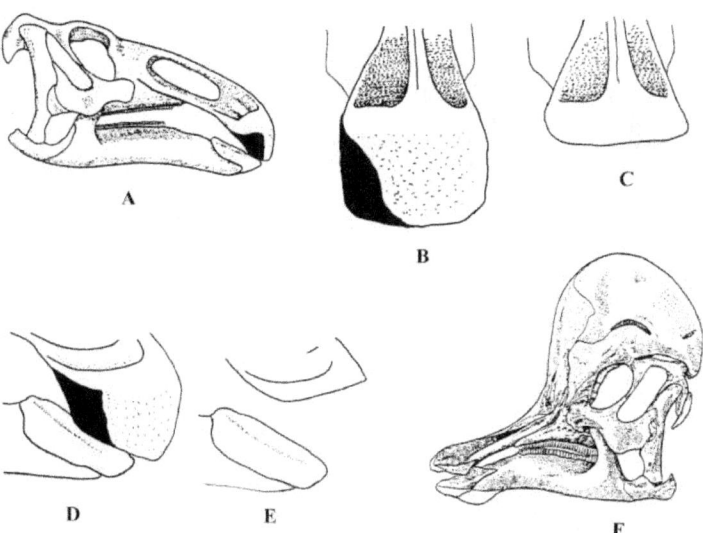

Figure 1) Skull of LACM 23502, a) Side view of skull; b) front view of the skull with rhamphotheca; c) front view of the skull without the rhamphotheca; d) side view of the beak with rhamphotheca; e) side view of the beak without the rhamphotheca; e) *Corythosaurus casuarus* showing how small the 'bill' is from side view (after Brown, 1914).

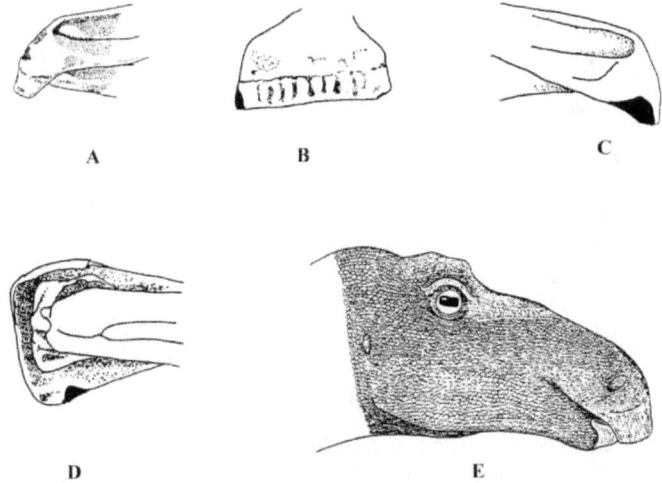

Figure 2) Senckenburg Museum skull, a) left side of bill showing the rhamphotheca; b) front view of the rhamphotheca; c) right side of bill showing the rhamphotheca; d) ventral view of the bill showing how wide the bill is; e) side view of a fleshed out head with rhamphotheca.

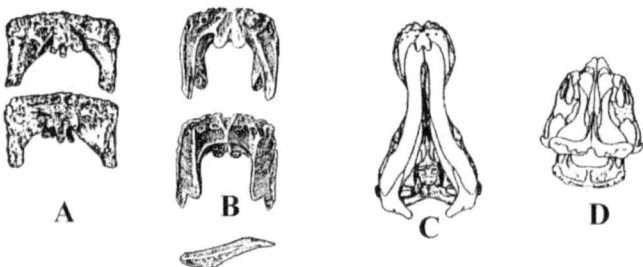

Figure 3) Predentaries, a) *Tsintaosaurus spinorhinus* in dorsal and ventral view (after Young, 1958); b) *Prosaurolophus blackfeetensis*, dorsal, ventral and lateral view; c) Ventral view of skull with dentary in place showing the squared area of the predentary; d) front view of *Prosaurolophus blackfeetensis* (after Horner, 1992).

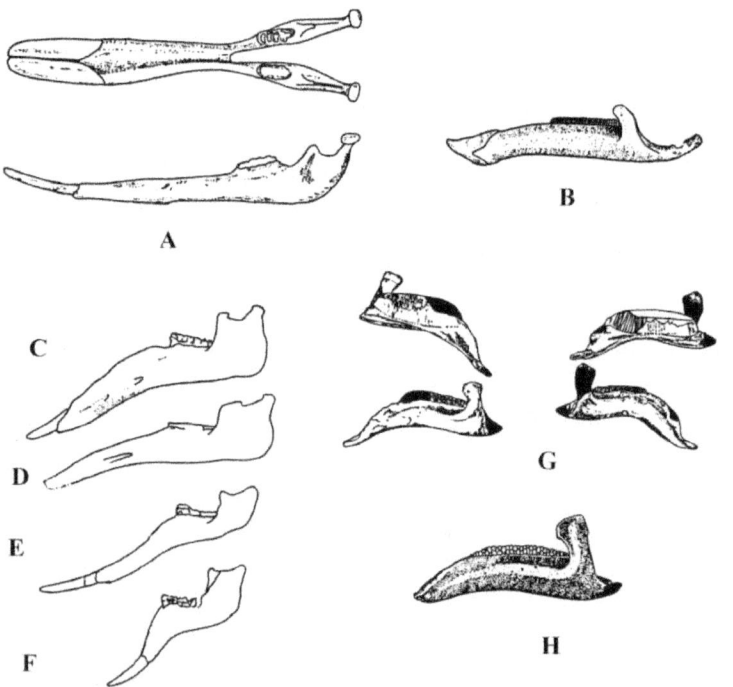

Figure 4)
Lower jaws of gomphotherers and hadrosaurs, a) Long jawed gomphothere *Trilophodon (Amebelodon) fricki*; b) A long jawed hadrosaur *Anatotitan copei*; c) Short jawed gomphotheres, *Trilophodon giganteus*; d) *Trilophodon lulli*; e) *Rhynchotherium edense*; f) *Rhynchotherium hondurensis*: g) *Tsintaosaurus spinorhinus* lateral view of left and right dentaries' (after Young, 1958); h) *Prosaurolophus blackfeetensis* lower jaw in lateral view (after Horner, 1992).

Bibliography

Brown, B. B., 1914, *Corythosaurus casuarius*, a new crested dinosaur from the Belly River Cretaceous, with provisional classification of the Family Trachodontidae: Bulletin of the American Museum of Natural History, v. 33, p. 559-565.

Horner, J. R., 1992, Cranial morphology of *Prosaurolophus* (Ornithischia: Hadrosauridae) with descriptions of two new hadrosaurid species and an evaluation of hadrosaurid phylogenetic relationships: Museum of the Rockies Occasional Paper n. 2, p. 1-119.

Morris, W. J., 1970, Hadrosaurian Dinosaur bills-Morphology and Function: Los Angeles County Museum, Contributions in Science, v. 193, p. 1-14.

Young, C.-C., 1958, The Dinosaurian Remains of Laiyang, Shantung: Palaeontologia Sincia, Whole Number 142, new series C, n. 16, p. 1-138.

PreHistoric Times

#54 June/July 2002

Paleoartist
Michael Skrepnick

DINOTOPIA
James Gurney

Marx Dinosaur
Playsets Part II

& much more

U.S. $5.95 • Canada $6.95

08

0 56698 94980 0

Ford, T. L., 2002, How to Draw Dinosaurs. Mitten, mitten, did hadrosaurs have mittens? Prehistoric Times, n. 54, p. 14-15

Chapter 8

Mitten, Mitten, did hadrosaurs have Mittens?

As promised last issues, this time I discuss whether or not hadrosaurs (and possibly other ornithopods) had mittens for hands (front toes covered by skin) or if there consisted of free toes. There are less than 6 well-mummified hadrosaur skeletons (at least that have currently been described). One *Edmontosaurus* (editors not, I now believe *Edmontosaurus*, *Anatosaurus* and *Anatotitan* are all valid genera and the *Edmontosaurus* specimens discussed in this article are *Anatosaurus*) is on display at the American Museum of Natural History, another at the Senckenburg Museum in Germany, I believe one was sent to Europe but the ship was sunk during WWII (the specimen may in fact still be intact if it was well packaged), two new *Brachycephalosaurus* specimens from Montana (in the Malta Museum, Montana) and another new *Edmontosaurus* for sale.

The *Edmontosaurus* at the AMNH is behind a glass and hard to see from all angles. Also as I mentioned last issue the cast of the Senckenburg *Edmontosaurus* was in Tucson during the Rock/Fossil/Gem Show and I was able to study it. The cast is for sale by Glen Rockers (PaleoSearch) and he also had the cast of one 'mitt' and he was very gracious to allow me to study. One mitt has the skin draped over the metacarpals on both sides (similar to when you have a plastic bag and suck out all the air), the other has the anterior skin draped over the metacarpals and the posterior side is budging (or seems to have had a pad). Both hands have the toes covered by skin with the 4th toe separate from the mitt. Which one is correct? What does the ichnology suggest? First we'll look at the structure of the metacarpals. Osborn (1912) originally thought the toes could be spread apart and were webbed which allowed the animal to use the hands like paddles for swimming. The toes look more like ungulates than reptiles, meaning the three metacarpals I-III were tightly bound and could not be separated like Osborn believed (Figure 1).

The ichnology shows that the front does were in a very shallow 'V' and shows no 'pad' (I've reported this in Prehistoric Times, 35). Both mitts had the soft tissue eroded away. The one with the 'pad' was possibly filled in with matrix and may have expanded the mitt. In life the skin enclosed the first two toes, and the ungual/claw was completely covered ventrally, but showed dorsally. Digit 3 was not completely included in the mitt in the Senckenberg specimen and may be due to an artifact of preservation or preparation because the AMNH specimen has the first three toes covered with in the mitt. In both specimens digit 4 was free of the mitt. Digit 4 may have been totally included in skin or complete separate (Figure 2).

Conclusion? It seems that hadrosaurs and the larger 'quadrapedal' Iguanodontids had mitts (as Mark Hallet has shown for over 15 years. He wrote an article for one of the Dinosaurs Past and Present Volume, Hallet, M., 1987, Bringing Dinosaurs to life: In: Dinosaurs Past and Present, v. 1, p. 96-113. Credit to where credit is due, I always say).

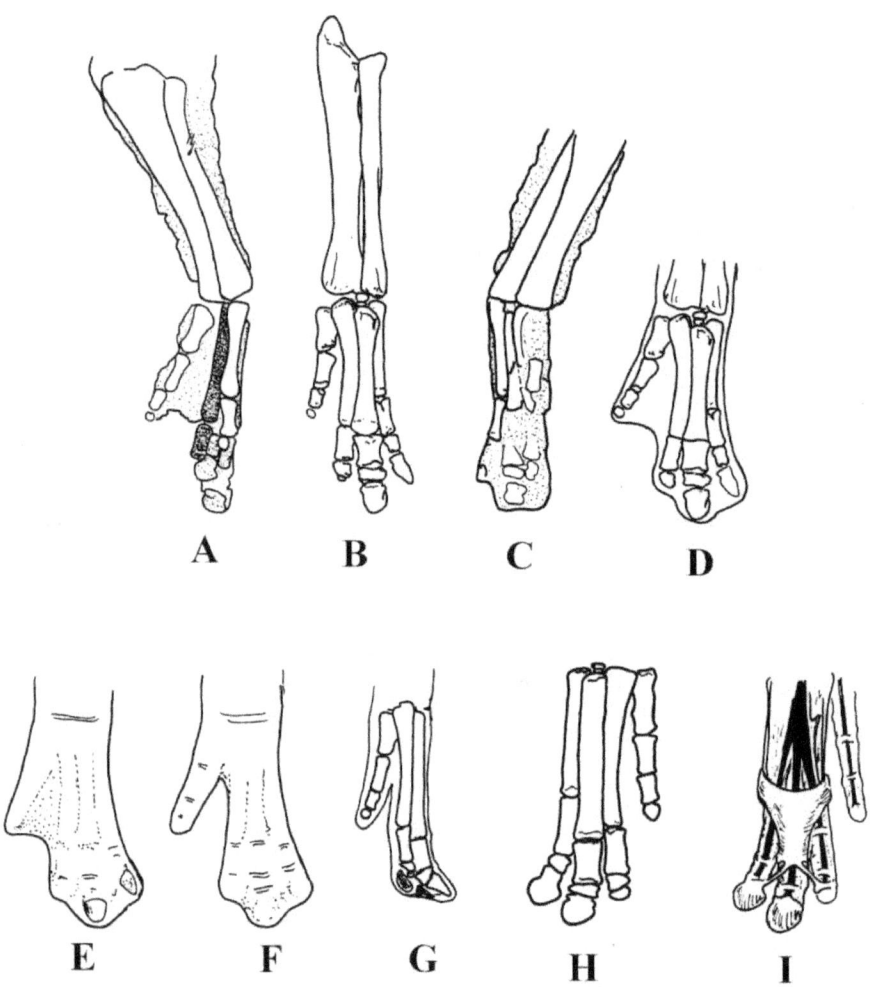

Figure 1). *Edmontosaurus annectens* (AMNH 5060), A-F; *Maiasaura peeblesorum* G-I. A) lower right arm with skin, B) lower right arm skeleton, C) Left lower arm, D) hand skeleton showing skin outline, E) hand showing digit 4 incased in skin, F) hand showing digit 4 'loose', G) lateral view of hand showing 'pads' on the ventral edge of toes, H) hand skeleton, I) illustration showing tendons and ligaments of the hand (modified from Hallett, 1987).

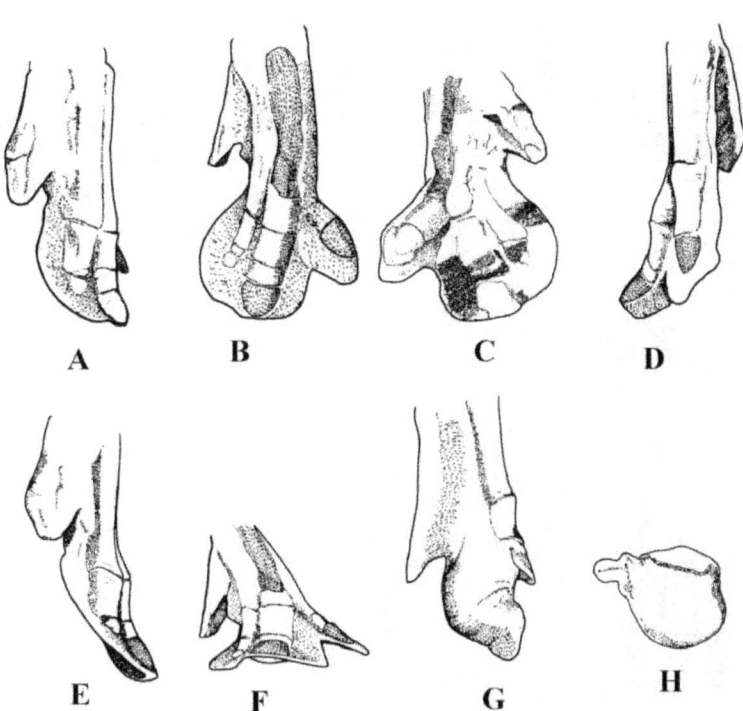

Figure
2). Senckenburg *Edmontosaurus annectens* mummy hands, A-F right hand and G-H left hand; A) outside view, B) dorsal view with unguals dark, C) ventral view, D) inside view, E) lower outside view, F), front view, G) ventral view of the 'expanded pad', H) front view.

Bibliography

Osborn, H. F., 1912, Integument of the Iguanodont dinosaur *Trachodon*: Memoirs of the American Museum of Natural History, new series, v. 1, p. 35-54.

PreHistoric Times

TIMES

#55 Aug/Sep 2002

BIG SUMMER 55TH ISSUE

Summer
Dino Getaways

Giant Croc!!
Artist **Gary Staab**

Paleontologist
James Kirkland

How To: Computer Generating Dinos

Sinclair Oil
in the 60s

Megalosaurs
Ichthyosaurs
& Ancient Bison

and much more

U.S. $5.95 • Canada $6.95

Ford, T. L., 2002, How to Draw Dinosaurs. Ankylosaurs, the checkerboard dinosaur: Prehistoric Times, n. 55, p. 14-15.

Chapter 9

Ankylosaurus, the checkerboard dinosaur

This issue's article covers a subject that I've been researching on for many years now, just what did the armor of *Ankylosaurus* look like. My work began when Jim Kirkland visited the San Diego Natural History Museum in the late 90's and he told me I had to work on the 'nodosaur' from San Diego. My job situation changed so I was able to prepare and describe the San Diego Ankylosaur, *Aletopelta coombsi* (We did this on a joint paper and incidentally this is the FIRST dinosaur named from California). As it turns out it's not a nodosaur as was originally thought, but an ankylosaur (a conclusion Jim and I came to independently from each other). In researching on *Aletopelta* I visited museums all over the U.S. and Canada. The type material of *Ankylosaurus* is at the American Museum of Natural History and consists of a fragmentary skull, fragmentary skeleton and armor from two individuals. I've already wrote about the skeleton of *Ankylosaurus* way back in Prehistoric Times number 28/29, p. 14-15 but didn't go into the armor. I recently gave a talk at the Tate Museum in Casper, Wyoming on this subject and this article is derived from that talk.

The armor of ankylosaurs varies and when *Ankylosaurus* was first described by Brown (1908) he believed it was similar to the only really good armored dinosaur known at that time, *Stegosaurus* (Figure 1a-b). It wasn't until more ankylosaurs were found that this interpretation was found to be incorrect. He also figured the armor in several rows with a single row down the midline. This is incorrect and I've commented on this fact before (Prehistoric Times, number 36, p. 14-15). The primitive condition of armor is in a double row down the middle of the back, or on either side of the neural arches. Brown described the *Ankylosaurus* as a heavily armored dinosaurs (Figure 1c). George Olshevsky made an ankylosaur family tree chart (1979) and figured a few dozen ankylosaurs (which many of his interpretations are still used today) and at the time was the best pretrial of ankylosaurs (Figure 2d). Ken Carpenter (1982) figured *Euoplocephalus* using the mummified skeleton of *Scolosaurus cutleri* and this has been the major view of *Euoplocephalus* and *Ankylosaurus,* ever since. I've commented on *Euoplocephalus* (Prehistoric Times number 45, p. 14-15) and won't go into that here. So, who is correct? Brown? Olshevsky? Or Carpenter?

In some ways they are all correct, and I'll go into that shortly. As I studied the material at the American Museum I noticed that the armor of *Ankylosaurus* wasn't oval with a ridge down the middle like *Euoplocephalus*, but more square with a ridge near the edge of one of its sides (Figure 2, 3). I first believed that this ridge would be pointing away from the midline, but that quickly changed when I ran across a piece of cervical bony ring with the armor in place. As I mentioned before, ankylosaurs have at least two cervical bony rings with 3 sets of armor (scutes). I've named the three sets as the median (middle), primary, and secondary (last, lower). The secondary scutes match the lower scute on the skull so it's easy to place the position of that scute. The primary scute has the ridge on the upper edge not the lower (as I thought would have) and the median scutes would have been the same (figure 4). Unfortunately the rest of the armor's placement is unknown. There is a similar loose piece of square armor that was placed on the body. There is a large long scute that is similar to the tertiary (lower than the secondary) that sat over the shoulder as seen in *Euoplocephalus*.

Trying to figure out the rest of the armor is difficult but not impossible. There are similar ankylosaurs with the same type of armor as *Ankylosaurus*. I've named an ankylosaur from New Mexico *Glyptodontopelta mimus* (I won't tell you what some paleontologist think of my mental state about naming dinosaurs on their armor). *Glyptodontopelta* has large square scutes with and pieces of the armor of the pelvic area. The pelvic armor is similar to *Stegopelta* and *Alectopelta* (which I put into the subfamily of Stegopeltinae). The armor is in irregular shapes that abut against one another (figure 5). The pelvis was a very solid piece of the skeleton and armor 'shield' would have been a natural extension of that. If all four were similar than it wouldn't be out of the ordinary that *Ankylosaurus* also had a pelvic 'shield'. When I visited the museum in Glen Rock, Wyoming, Sean Smith showed me some armor that they had been collecting. I've heard about this a few years ago and I was finally able to study it. Only a few pieces of the skeleton were found but lots of armor. The armor is from the middle of the back to the pelvis. There is one piece of armor that is square and has a ridge near the edge. This might belong to *Ankylosaurus* and if it does it too had a pelvic 'shield' like *Stegopleta* and *Glyptodontopelta*. The tail would have triangular scutes and of course a large tail club.

So, the body would have had a 'checker-board' type of armor, and did not look like *Euoplocephalus* or how Brown depicted it (Figure 6). But was Brown right? That'll be a topic for another time.

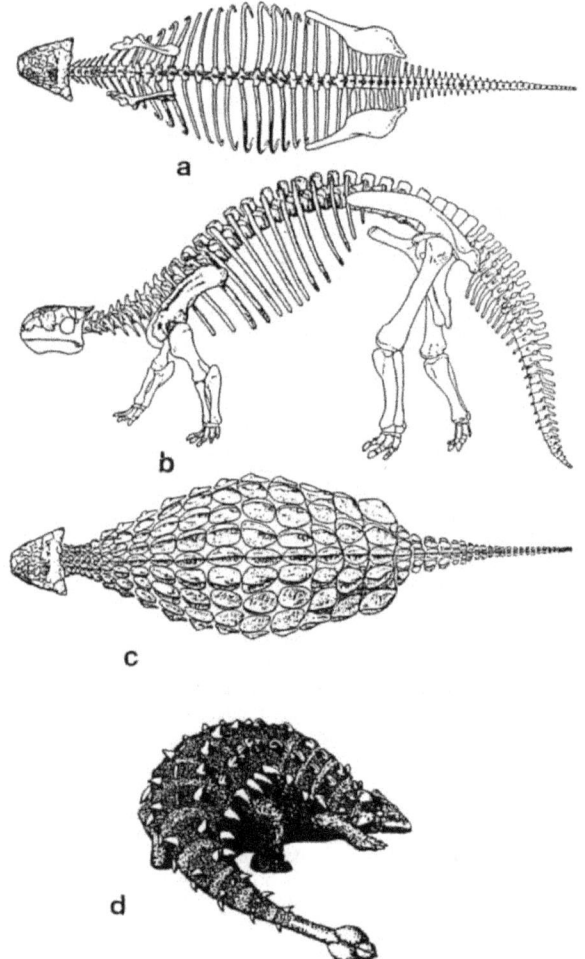

Figure 1). *Ankylosaurus magniventris* after Brown (1908); (a) Side view of the skeleton; (b) dorsal view of the skeleton; (c) dorsal view of the skeleton with armor, (d) George Olshevsky's reconstruction of *Ankylosaurus*.

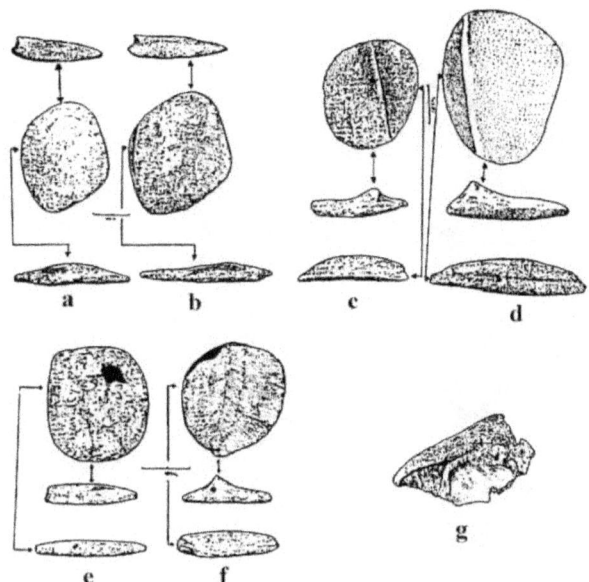

Figure 2) AMNH 5895 scutes. (a) Pectoral scute in anterior, dorsal and lateral view, (b) pectoral scute in anterior, dorsal and lateral view, (c) tertiary (?) scute in dorsal, anterior and lateral view, (d), pectoral scute in dorsal, anterior and lateral view, (e) pectoral scute in dorsal, anterior and lateral view, (f) pectoral scute (?) in dorsal, anterior and lateral view, (g) tertiary pectoral scute in oblique lateral view.

63

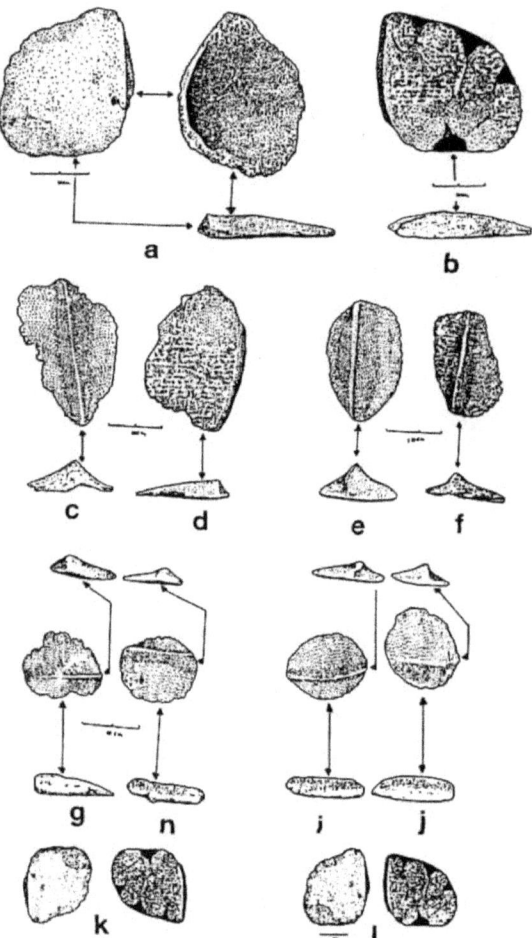

Figure 3) AMNH 5214 scutes. (a) Pectoral scute in dorsal, ventral and anterior view, (b) pectoral scute in dorsal and lateral view, (c) tertiary (?) scute in dorsal and anterior view, (d) tertiary scute in dorsal and anterior view, (e) tertiary (?) scute in dorsal and anterior view, (f) tertiary (?) scute in dorsal and anterior view, (g) tertiary body scute anterior, dorsal and lateral view, (h) tertiary body scute anterior, dorsal and lateral view, (i) tertiary body scute anterior, dorsal and lateral view, (j) tertiary body scute anterior, dorsal and lateral view, (k) pectoral scutes with ridge ventral, (l) pectoral scutes with ridge dorsally placed.

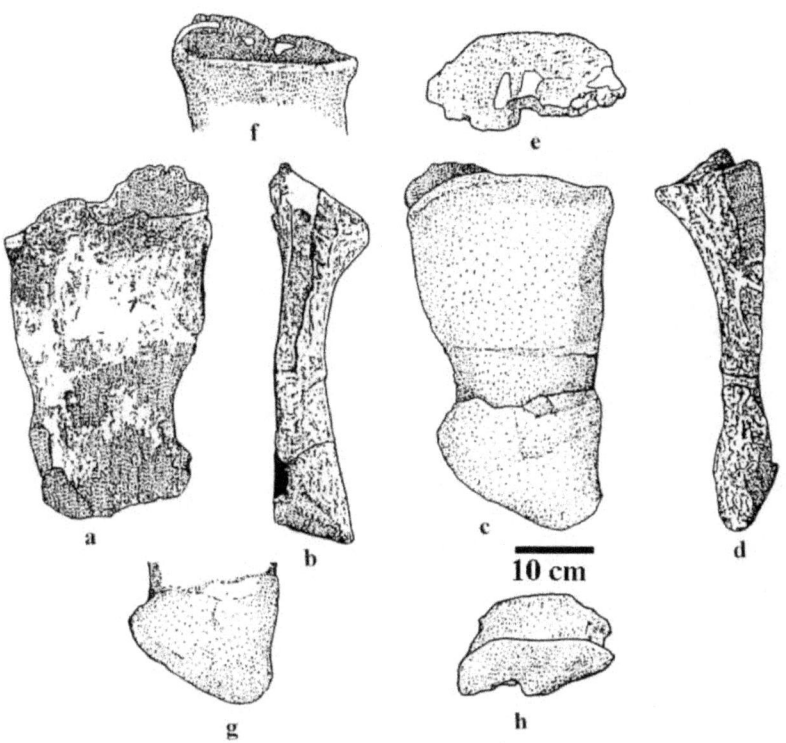

Figure 4) AMNH 5895
cervical ring half in (a) inside view, (b) posterior, (c) lateral, (d), anterior, (e) close up on the upper scute ridge, (f) dorsal view, (g) close up on lateral scute, (h) ventral view.

65

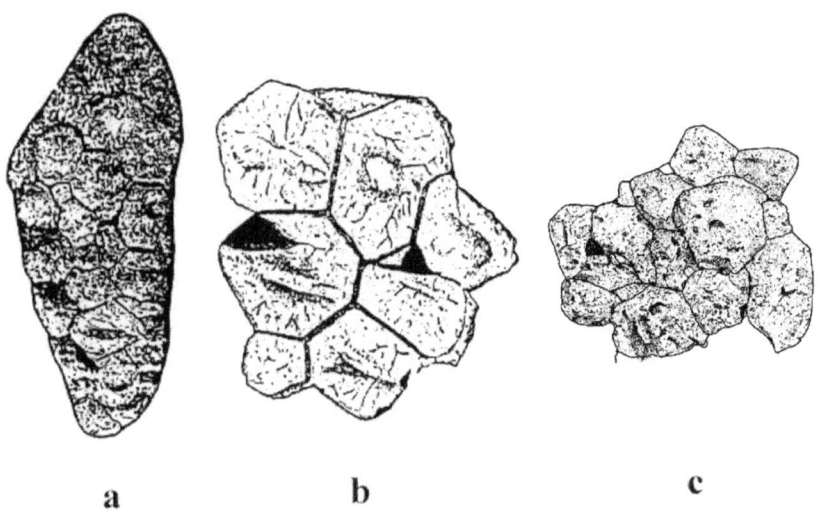

a   b   c

Figure

5) pelvic armor of ankylosaurids; a), *Stegopelta landerensis* (FMNH UR88); b) *Glyptodontopelta mimus* USNM 8610; c) *Aletopelta coombsi* (SDNHM 33909) in oblique view.

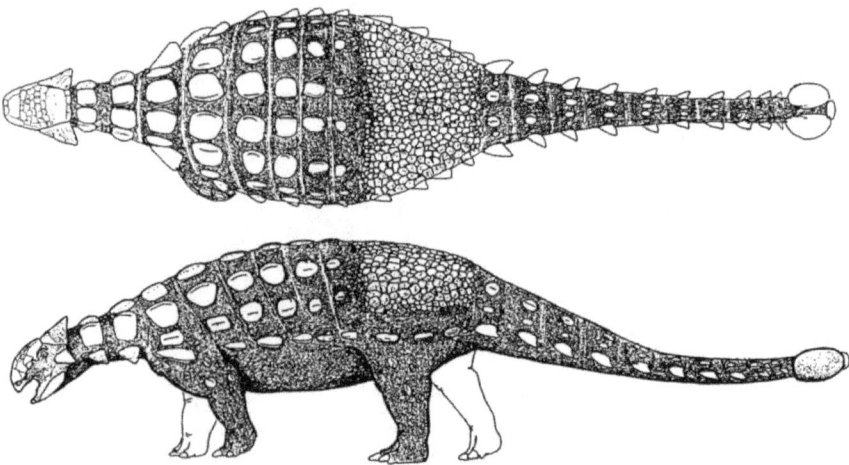

Figure 6) Dorsal and lateral view of the skeleton and dorsal and lateral view of a restored *Ankylosaurus. magniventris.*

Bibliography

Brown, B. B., 1908, The Ankylosauridae, a new family of armored dinosaurs from the Upper Cretaceous: Bulletin of the American Museum of Natural History, v. 24, p. 187-201.

Carpenter, K., 1982, Skeletal and dermal armor reconstruction of *Euoplocephalus tutus* (Ornithischia: Ankylosauridae) from the Late Cretaceous Oldman Formation of Alberta: Canadian Journal of Earth Science, v. 19, p. 689-697.

Ford, T. L., 2000, A review of ankylosaur osteoderms from New Mexico and a preliminary review of ankylosaur armor: In: Dinosaurs of New Mexico, edited by Lucas, S. G., and Heckert, A. B., New Mexico Museum of Natural History and Science Bulletin n. 17, p. 157-176.

Ford, T. L., 2002, A new look at the armor of *Ankylosaurus*: In: Tate 2002, Wyoming in the Age of Dinosaurs, Creatures, Environments and Extinctions, p. 15.

Ford, T. L., 2003 (for 2002), A new look at the armor of *Ankylosaurus*, just what did it look like?: In: The Mesozoic in Wyoming, Tate 2002, Casper College, p 48-67.

Ford, T. L., and Kirkland, J. I., 2001, Carlsbad ankylosaur (Ornithischia, Ankylosauria): an ankylosaurid and not a nodosaurid: In: The Armored Dinosaurs, edited by Carpenter, K., Indiana University Press, p. 239-260.

PREHISTORIC TIMES

#56 Oct/Nov 200

The PT Interview:
Ray Harryhausen

The prehistoric sculptures of
Sean Cooper

Making the Raptors for
Jurassic Park

U.S. $5.95 • Canada $6.95

Ford, T. L., 2002, How to Draw Dinosaurs. What did the jaws of spinosaurids look like? Prehistoric Times, n. 56, p. 14-15.

Chapter 10

What did the jaws of spinosaurids look like?

Spinosaurids are a fascinating and bizarre group of theropods that sparks the imagination. Their long, short skulls and sails leave some to believe they swam and eat fish. They are not a well-known group; unfortunately the type of *Spinosaurus* was destroyed by allied bombers during WWII so only new findings will help solve the mystery of its head. Thanks to recent discoveries by Sereno et al, (*Suchomimus*), and Charig & Milner (*Baryonyx*) the head of spinosaurids are better known. There are at least three families of spinosaurids, Irriatoridae, Spinosauridae and Baryonyxidae. There are at least two types of *Spinosaurus's*; *S. aegyptiacus* (the type) and *Spinosaurus maroccanus*. The latter was first described by Russell on a tip of dentary, part of a maxilla, cervical vertebrae and a neural arch. A better specimen, a toothless snout, was described by Taquet and Russell and this specimens shows a different morphology than *S. aegyptiacus*. By using *Suchomimus* and *Baryonyx*, then relating that to *Spinosaurus*, a skull can be theorized.

The skull of *S. aegyptiacus* would have been shorter and stockier while the skull of *S. maroccanus* was longer. I won't be going into the Irriatoridae because *Irritator* is being described by Sues and I'll wait for this paper to come out.

There are at least 2 genera of Baryonyxidae, *Baryonyx walkeri* and *Suchomimus tenerensis* and I will be using *Suchomimus* for the article along with *Spinosaurus maroccanus*.

On a side note; I fondly remember the talk given by Alan Charig on *Baryonyx* during the 1986 Dinosaur Systematic Symposium held at the Tyrrell Museum of Paleontology (Held long before it became the Royal Tyrrell Museum of Paleontology). First he gave a summary of the find, age, etc. Then he said if any of you know what it is, please let me know (or some such), then he showed a slide of the skull. The audience gasped and we were all aw struck. Rauischid, crocodilian, theropod, were all mentioned. None of us at the time knew just how strange this theropod was.

What I will comment on are already described specimens. *Suchomimus* is the best-known Baryonyxid spinosaurid (though a good monograph on it hasn't been done yet like there is for *Baryonyx*. *Suchomimus* has the longest skull of any known theropod and has been strongly suggested that it lived a life like a crocodilian. I disagree with this theory and believe it was more terrestrial than aquatic because the body does not support an aquatic lifestyle. The teeth have been, at times, incorrectly interpreted as being interlocking like a crocodilian. The jaws of spinosaurids are typical of any theropod.

The lower jaw fits into the upper jaw and the upper teeth overhang (not interlock) the lower jaw. The dentary is laterally compressed and was very very thin. (Fig 1, 2). The skull was long and low and the dentary extremely narrow, but fit perfectly into the upper jaw. The dentary teeth fit inside the upper jaw, which is typical of theropods (See PT, 42, p. 49-50, also How to Draw Dinosaurs volume 1, chapter 5, p. 26-30). The premaxillary teeth are large (about the largest in the skull) with the front of the maxillary teeth of nearly equal size.

Baryonyx had a shorter skull than *Suchomimus* and *Spinosaurus aegyptiacus* had a shorter skull than *Spinosaurus maroccanus* (Figure 3). Oliver W. M. Rauhut has a new theory about the eating habits of spinosaurids. His study shows that the skull couldn't resist lateral forces (i.e. side to side action as other theropods used) and also because the skull is so long had less resistant against vertical forces. He suggests that the because of the long skull it could close it's mouth more rapidly which aided in capturing smaller prey. He does not mention whether or not it ate fish in the abstract (that is all that was published) but does not rule out the possibility (pers. comm..). Also in question is whether the fish and *Iguanodon* material found with *Baryonyx* was actually in the stomach region of the skeleton.

On a similar not, Theargten Lingham-Soliar has written a paper on the caniform premaxillary teeth of the mosasaur *Goronyosaurus*. In it he states that this mosasaur, with large premaxillary teeth, closed it's jaws rapidly, but lost some muscle strength for crushing, just like in spinosaurids. It looks like spinosaurids caught smaller prey with rapidly closing jaws, and not larger prey. This would also suggest they didn't use their heads to feed on carrion as they couldn't move their heads side to side to rip off pieces of meat. This would mean *Suchomimus* didn't take down sauropods, but probably left that for *Carcharodontosaurus*.

There are new specimens of *Spinosaurus* being described (2 by Angela Milner, 2 front halves of skulls which look more like phytosaurs than theropods, and possibly belonging to *S. maroccanus*), and new specimen from Morocco held at an Italian Museum. A cast of the front half of the skull was on display by Glen Rockers at the Tucson Rock Show and he told me hopefully the back half will be casted and a complete skull (or as complete as possible) should be on display next year. I certainly hope so, that's going to be a huge skull! This skull looks more like *Baryonyx* then *Spinosaurus*. When they are all described I'll write them up for PT.

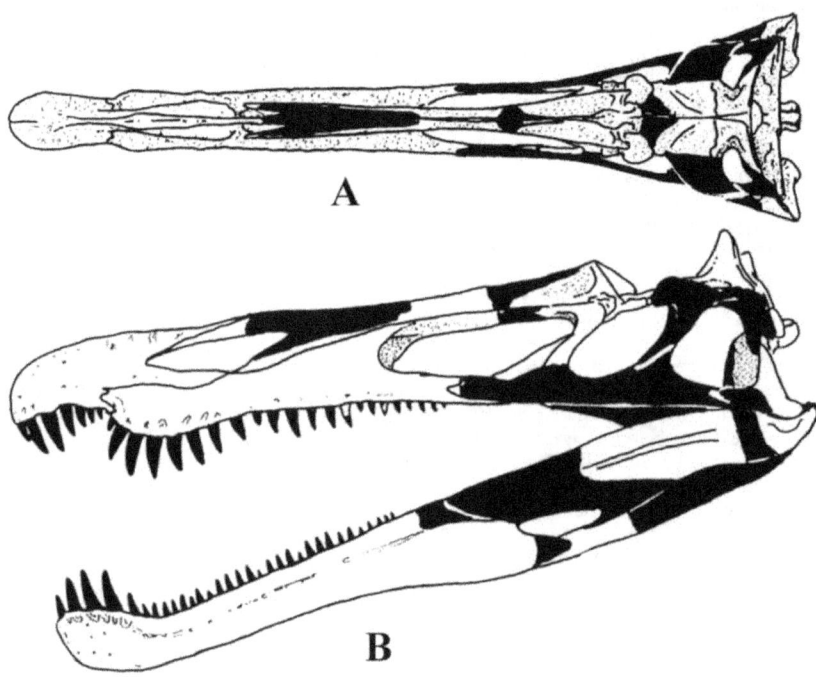

Figure 1). Skull of *Suchomimus*; A) top view of skull; B) side view of skull (modified from Sereno, et al., 1998).

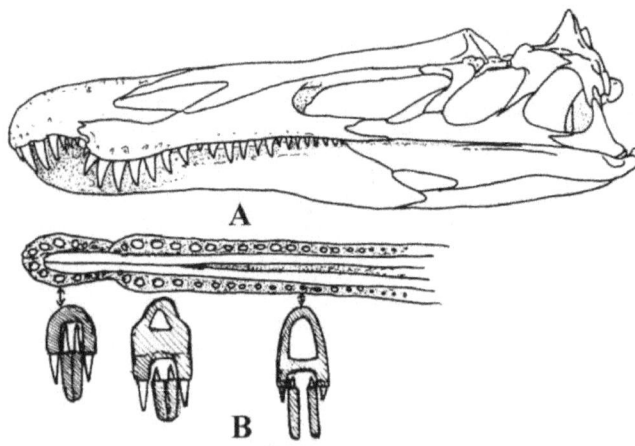

70

Figure 2). Skull of *Suchomimus*; A), side view of skull with closed mouth; B), ventral view of skull showing the dentary articulation with the upper jaw. Notice how thing the lower jaw is (modified from Sereno, et al., 1998).

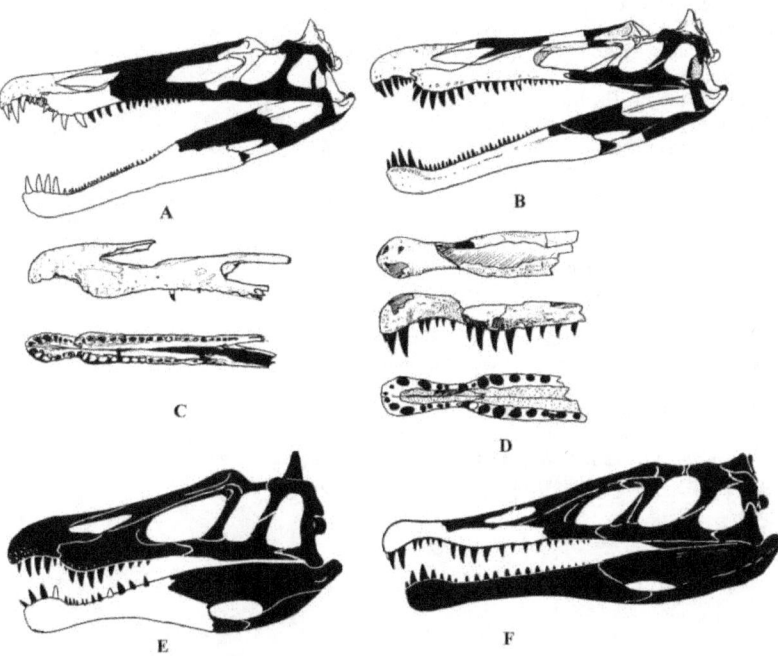

Figure 3). Skulls of Spinosaurids; A) *Baryonyx* (after Charig, & Milner, 1997); B) *Suchomimus* (modified from Sereno, et al., 1998); C) premaxilla & maxilla of *Suchomimus* (after from Sereno, et al., 1998); D) premaxilla and maxilla of *Spinosaurus maroccanus* (after Taquet & Russell, 1998); E) *Spinosaurus aegyptiacus* (modified from Stromer, 1917); F) *Spinosaurus maroccanu* (modified Taquet & Russell, 1998).

Bibliography

Charig, A. J., and Milner, A. C., 1986, *Baryonyx*, a remarkable new theropod dinosaur: Nature, v. 324, p. 359-361.

Charig, A. J., and Milner, A. C., 1997, *Baryonyx walkeri*, a fish-eating dinosaur from the Wealden of Surrey: Bulletin of The Natural History Museum, Geology Series, v. 53, n. 1, p. 11-70.

Sereno, P. C., Beck, A. L., Dutheil, D. B., Gado, B., Larsson, H. C. E., Lyon, G. H., Marcot, J. D., Rauhut, O. W. M., Sadleir, R. W., Sidor, C. A., Varricchio, D. D., Willson, G. P., and Wilson, J. A., 1998, A Long-Snouted Predatory Dinosaur from Africa and the Evolution of Spinosaurids: Science, v. 282, p. 1298-1302.

Stromer, E., 1915, Ergebnisse der Forschungsreisen Prof. E. Stromers in den Wusten Agyptens. II. Wirbeltier-Reste der Baharije-Stufe (unterstes Cenoman). 3. Das Original des Theropoden *Spinosaurus aegyptiacus*: Abhandlungen der Koniglich Bayerischen Akademie der Wissenschaften Mathematisch-physikalische Klasse 28, band 3, p. 3-32.

Taquet, P., and Russell, D. A., 1998, New data on spinosaurid dinosaurs from the Early Cretaceous of Sahara: Compte rendu hebdomadaire des seances de l'Academie des Sciences Paris, Siences de la terre et des plantes, v. 327, p. 347-353.

A new look for the head of *Tanystropheus*

At his years SVP (2002) I'm giving a poster on a new interpretation of the head of *Tanystropheus* and I will summarize my poster here. The head of *Tanystropheus* has been illustrated conservatively with overhanging teeth and straight jaw line. Looking at photo's of the skulls of the different Tanystropheids show that the front teeth actually interlocked and were conical in shape looking more like a pleisosaur or pterosaur (This all started when Charlie MaGrady asked me for some information on *Tanystropheus*) (Figure 1 and 2). The lifestyle has been interpreted as terrestrial animals that stuck their heads in the water to catch fish, or as juveniles catching insects. Some specimens have been found with beleminte hooklets so we know they were swimming around the ocean. The nose has also been misinterpreted as being on the sides of the skull and closer to the premaxilla, similar to lizards. My interpretation shows that the nasals were dorsal (or higher) on the skull like that of mosasaurs, and a dorsally placed nasal is a good indication of an aquatic lifestyle.

I believe *Tanystropheus* was a completely aquatic animal that could venture onto land if it wanted to. The neck is long (11 cervical vertebrae, long, low, now neural arch and hollow) and it had a 'stiff' neck (Figure 3). As reported in the last issue of PT elasmosaurs also had a 'stiff' neck even though it had dozens of vertebrae. *Tanystorpheus* and elasmosaurs had the same type of neck morphology though they achieved it via different ways. *Tanystropheus* has long cervical ribs that cover two to three vertebrae and formed bundles. It is a misinterpretation that this would make the movement of the neck impossible. The cervical ribs could bend and was attachment areas for muscles. Several specimens indicate the bending of cervical rib (figure 4).

Looking at articulated or semi articulated specimens indicates the behavior of the animal. The posterior part of the neck is more mobile than the rest of the neck, and it swam using its mobile body and tail. Sauropterygians on the other hand have stiff bodies and large paddles. I believe *Tanystropheus* was a more graceful slow swimming animal than sauropteyrgians. They used their hind limbs (even though they lacked paddles) and tail for propulsion for swimming, and maneuvered with the front limbs and bodies.

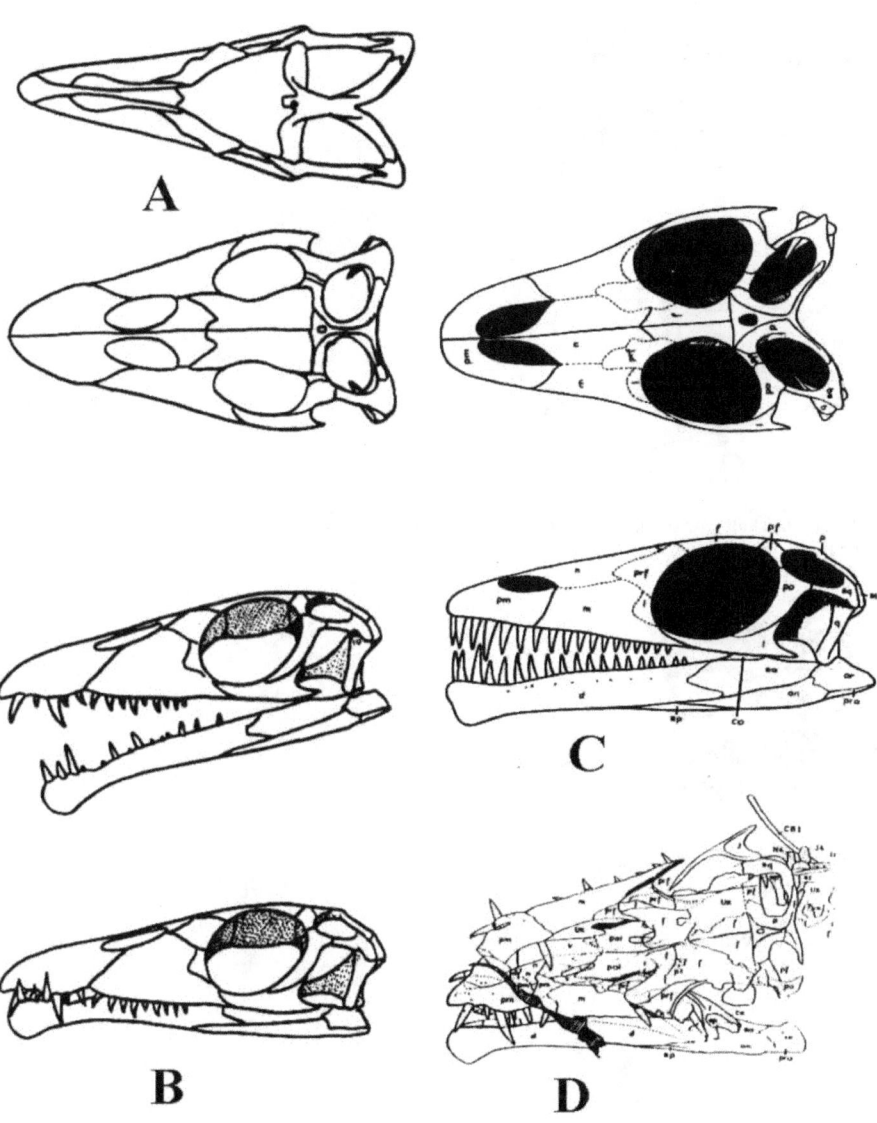

Figure 1). The skull of *Tanystropheus longobardicus* (examplar q of Wild, 1973); A) skull of a mosasaur showing the dorsally placed nostrils; B) my interpretation of the skull, top, side with mouth open, with mouth closed; C) Wild's original interpretation; D) showing the interlocking teeth.

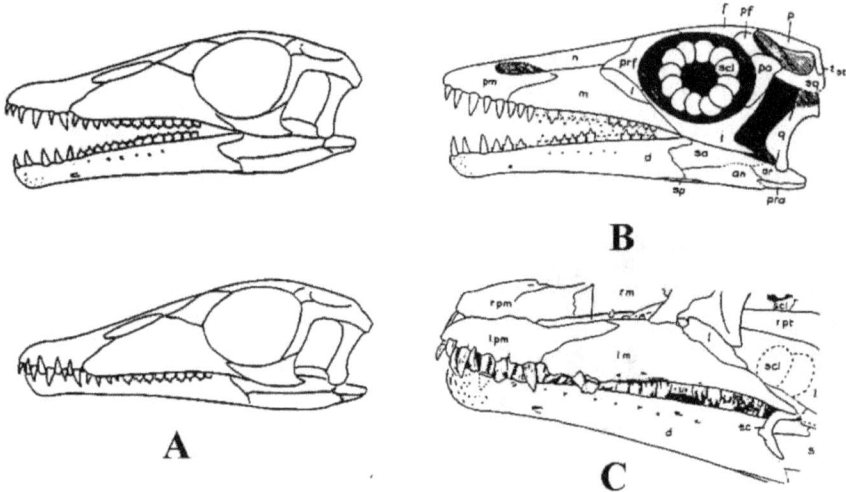

Figure
2). The skull of *Tanystropheus meridensis*; A) my interpretation of the skull with mouth open, and closed; B) Wild's
interpretation; C) close up of the jaws to show the interlocking teeth.

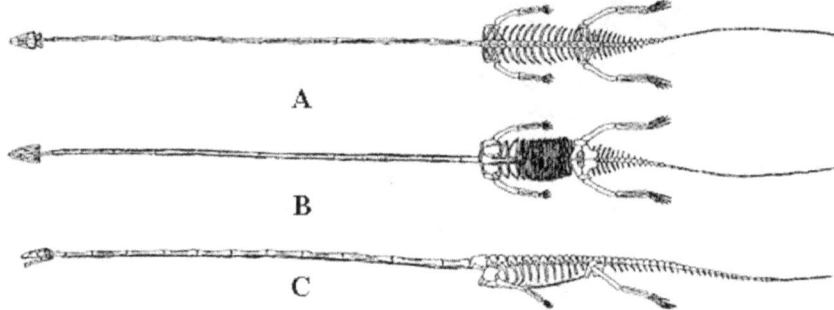

Figure 3). Skeleton of *Tanystropheus*: A) top view; B) ventral view; C) lateral view.

Figure 4). Skeleton of a young *Tanystropheus longobardicus* (examplar a of Wild, 1973) showing how it swam in water (that is my interpretation of the skeleton).

Bibliography

Ford, T. L., 2002, A new interpretation of the skull of *Tanystropheus*: Journal of Vertebrate Paleontology, v. 22, supplement to n. 3, Abstracts of Papers, Sixty-second annual Meeting of the Society of Vertebrate Paleontology, Sam Noble Oklahoma Museum of Natural History, University of Oklahoma, Norman Oklahoma, October 9-12, p. 53a.

Wild, P., 1973, Die Triasfauna der Tessiner Kalkalpen, XXIII. *Tanystropheus longopbardicus* (BASSANI): Schweizerische Palaontologische Abhandlungen Memoires suisses de Paleontologie, v. 95, p. 1-162.

PreHistoric Times

Dec/Jan 2003 #57

Sculptor Max Salas
Sea Scorpions
Parasaurolophus
Waterhouse Hawkins
and much more!

U.S. $5.95 • Canada $6.95

12

0 56698 94980 0

Ford, T. L., 2002-2003, How to Draw Dinosaurs. Winging it with dromaeosaurs: Prehistoric Times, n. 57, p. 14-15.
Chapter 11

Winging it with Dromaeosaurs

The theory that birds evolved from theropods has grained significant ground, although there are still those who believe otherwise. The discovery of several different lineages of 'feathered' theropods found in the famous Liaoning area has been a great boon for the evolution of birds theory. I must confess that a decade ago I was totally against this myself, but due to recent finds, I know that I'm wrong. Just about everyone, including me, scoffed at Greg Paul's feathered *Deinonychus* (1987), but he is constantly being shown that many of his theories are correct. What we'll be looking at in this article is the appearance of the wing (which should be the proper term).

The type of 'feathers' known so far from theropods are long hair like 'feathers' (proto-feathers?) that covered the body (that is until Stephen Czerkas book, Feathered Dinosaurs and the origin of flight). These hair like feathers have been found in the segnosaur *Beipiaosaurus*, compsognathid *Saurosauropteryx*, Dromaeosaurids, *Microraptor*, *Sinornithosaurus*, NGMC 91, and hundreds of birds. In *Sinosauropteryx* and NGMC 91 (an unnamed dromaeosaurid) the feathers start at the middle of the head and continue to the tip of the tail. Does this mean all feathered theropods had the feathers starting at the middle of the head? No. Not all birds have feathered heads, so it is not out of the question that not all 'feathered' theropods had 'feathered' heads. (Vultures lack feathered heads because they will stick their head deep into the body cavity of a carcass to feed). The 'wings' and legs of these theropods were covered in hair like feathers, some of which extended for several centimeters.

Before we get into the 'wing' itself we need to study a bird's wing and understand its anatomical composition. Watching a bird soar the front of the wing looks straight and one would believe that the skeletal wing is straight, but this is not true. The skeletal wing is bent (see figure 1).

The straight edge is the patagium, which consists of the Patagialis longus tendon (which attaches from the humeral head to the ulnare, which lies just behind the pollex or thumb, digit 1). Behind the tendon is the patagium, which is a tough band of tendinous tissues, which holds the quills of feathers. Below the skeletal wing is the postpatagium, which the flight feathers attach to. The feathers of the wing are in different layers. The first are the coverts; the first are the smaller marginal coverts, which has several 'rows', then lesser converts, the longer median secondary coverts, and the greater coverts. The 'thumb' has the alular quills. The flight feathers are the greater primary, primaries and secondaries. (all this was taken from Proctor and Lynch's book; Manual of Ornithology, and if you don't have it you should get it!).

Is this how a dromaeosaur wing looked like? If so did all dromaeosaurs have feathered wings? The answer to both is maybe. There is no evidence for a patagium in *Archaeopteryx* or dromaeosaurids (editors note, this is incorrect, Martin & Lim, 2005 have found a patagium in *Archaeopteryx*) (Figure 1, illustrations 3, 4), as there is no evidence for alular feathers. This would mean they lack the Patagialis longus tendon and the front of the wing would be 'thinner' than modern birds. It is important to note that the primary wing feathers attach to the second digit of the 'wing' in theropods, not the third. One prevailing theory that some theropods did have primary feathers though there is no physical evidence for feathers is a thick metacarpal II. Digit III was not held in the postpatagium and was free. I've seen several *Confuciusornis* specimens that will have digit III held perpendicular to digit II and UNDER the wing feathers. This is very important to remember, the wing feathers are 'above or dorsal) to the digits. This may mean that digit III was used in climbing and holding the animal in a tree. A climbing animal has long recurved claws and dromaeosaurs fit that bill. I won't go into whether or not dromaeosaurids or some theropods could or did climb trees other than to say that one day I hope to publish on this and when I do I'll do a PT article. It is unknown when the patagium started in the evolution of birds.

Many of the first birds lacked the alular feather but some are known to have it. The alular feathers helps in the maneuverability of the bird in flight and indicates a better flyer than those that lack it.

What about the wing feathers themselves? Did dromaeosaurs have convert feathers? We do know they had the long hair like feathers on the wing but so far no convert feathers have been found. But there are at least two specimens of a new genera of dromaeosaur that does have a-symmetrical flight feathers. *Cryptovolans pauli* (named after Greg Paul) and described by Czerkas, Zhang, Li and Li has long flight feathers on the wings, longer than *Archaeopteryx* (Figure 2). It also has the long hair like feathers, long claws which show some of the sheath, a large flat sternum (like *Confuciusornis*) and they believe *Cryptovolans* could fly. I also believe it could fly, but not in the way that modern birds can. We need to stop thinking of how modern birds fly and start thinking of how the first flyers could fly. *Cryptovolans* may have been a 'short' flyer. The tail is long and does show some evidence for feathers at the tip of the tail. These feathers are long and probably supported a small 'fan' of feathers. There is also evidence for feathers on the legs themselves. Not just the long hair like feathers as in NGMC 91, but long 'flight?' feathers.

Well, I won't go into that right now. There are other specimens waiting description that will answer that and when that happens I'll write that up for PT (can't wait!). *Cryptovolans* is the only theropod so far known with flight feathers. Did all Dromaeosaurs have flight feathers? We just don't know, but it is possible that some if not all of them did. NGMC 91 could have had all its flight feathers either not preserved or have been completely displaced (as Czerkas and Yuan theorize).

Czerkas et al also believe that manoraptorian non-avian theropods should no longer be considered non-avian theropods, but full-blown avian theropods or birds. This is controversial, and the way the book is written has offended some, though I do get the gist of the book and agree that this may be correct. Others are now and have in the past mentioned this for specific groups of manoraptorians and I have heard, second hand that is, that Larry Martin has now conceded that manoraptorians are birds, though I'd like to see this in writing first.

The book also describes other theropods and birds. The really cool one is the one with the long third digit, *Scansoriopteryx heilmanni* (Czerkas and Yuan). This is a baby or possibly a hatchling; a fragmentary skull, and the majority of the skeleton. The 'wings' are nearly intact and are very long, especially as I mentioned before digit 3. The pes has a reversed digit 1 that sits on the ground like modern perching birds! There is also evidence for the hair-like feathers seen in other theropods, along the head, and wing fingers. There is a controversy though; another theropod that looks uncannily like *Scansoriopteryx* was described; *Epidendrosaurus ningchengensis* (electronic version of a printed paper that has a September publication date, a month after Czerkas et al's book). This is a more fragmentary animal, but does show many similarities, the main one being the long third digit. Both were printed in the same month, but *Scansoriopteryx* was published 'first' in written format, while *Epidendrosaurus* was done on an electronic version of a printed-paper. For now I believe *Scansoriopteryx* has printed priority over *Epidendrosaurus*. What the long third digit was used for is unknown. It is not as thin as the Aye-aye and was probably not used in the same manner (to get insects).

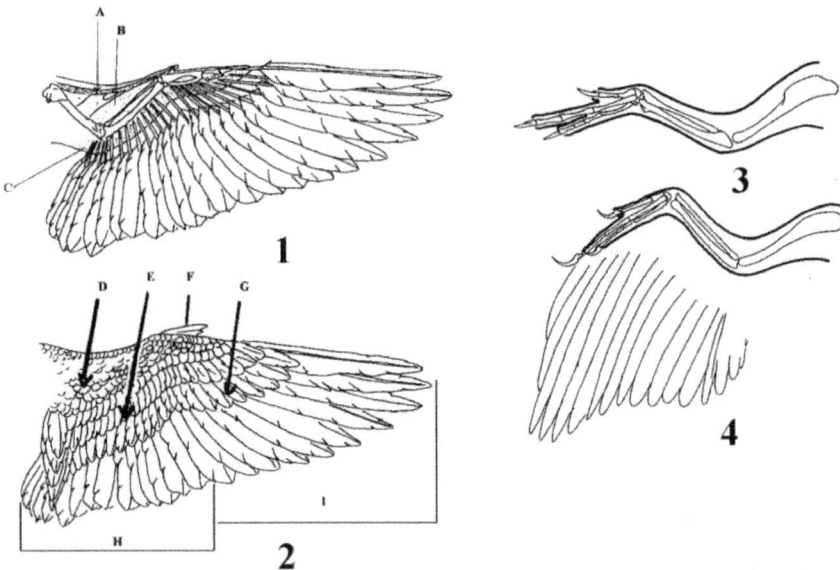

Figure 1. Wing of a Rock Dove (*Columba livia*). You can easily see the skeletal wing bones and how they are bent. 1) cutaway view of a wing; a) patagialis longus tendon; b) patagium; c) postpatagium. 2) feathers on the wing; d) marginal coverts; e) median secondary coverts and below that are the greater secondary coverts and above the median secondary are the lesser secondary coverts; f) alular quills; g) greater primary coverts; h) secondaries; i) primaries. 3) a wing of *Deinonychus* lacking the patagium and 4) wing of *Archaeopteryx* lacking the patagium (after Proctor & Lynch, 1993).

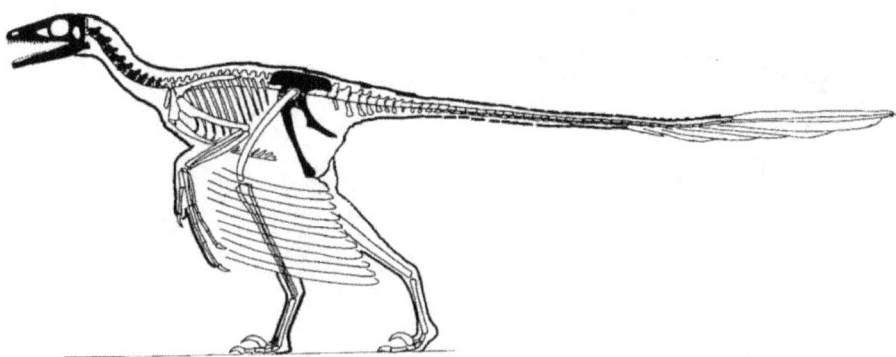

Figure 2) *Cryptovolans pauli* skeleton, modifed from Czerkas, et al., 2002.

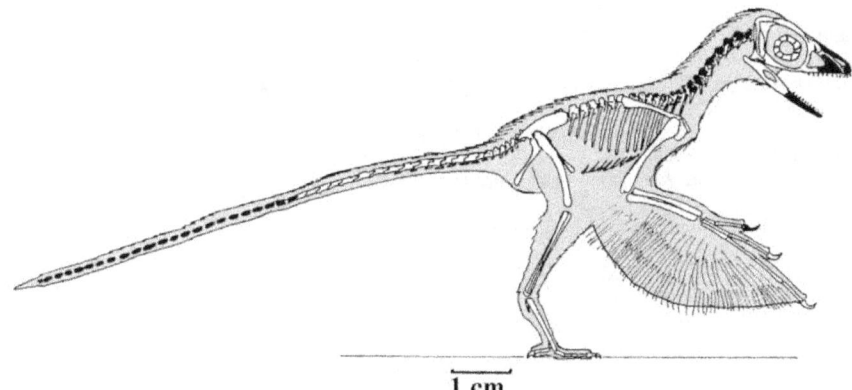

1 cm

Figure 3). *Scansoriopteryx heilmanni* skeleton, modified from Czerkas & Yuang, 2002.

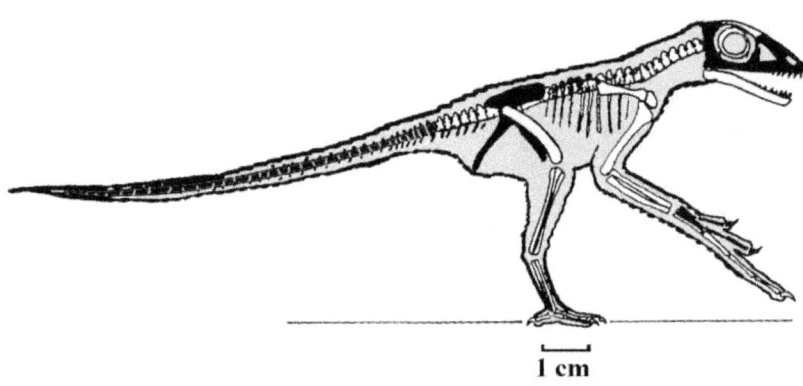

Figure
4) *Epidendrosaurus ningchengensis* skeleton, modified from Zhang, et al., 2002.

Bibliography

Czerkas, S. A., and Yuan, C., 2002, An arboreal maniraptoran from northeast China: In: Feathered Dinosaurs and the origin of flight, edited by Czerkas, S. J., The Dinosaur Museum Journal, v. 1, p. 63-95.

Czerkas, S. A., Zhang, D., Li, J., and Li, Y., 2002, Flying Dromaeosaurs: In: Feathered Dinosaurs and the origin of flight, edited by Czerkas, S. J., The Dinosaur Museum Journal, v. 1, p. 97-126.

Proctor, N. S., and Lynch, P. J., 1993, Manual of ornithology, avian structure & function. Yale University Press, 340pp.

Zhang, F., Zhou, X., Xu, X., and Wang, X., 2002, A juvenile coelurosaurian theropod from China indicates arboreal habits: Naturwissenschaften, v. 89, p. 394-398.

Last issue I commented that I'd show what the new *Irritator* skull looked like when the paper came out. The skull is a bit longer, the orbits a bit higher, and a strange looking animal, to say the least. I've put the snout of *Angaturama* on it because many believe they are one and the same animal, even the author of *Angaturama*, Kellner. (Figure 5). Sues, H.-D., Frey, E., Martill, D. M., and Scott, D. M., 2002, *Irritator challengeri*, a spinosaurid (Dinosauria: theropoda) from the Lower Cretaceous of Brazil: Journal of Vertebrate Paleontology, v. 22, n. 3, p. 535-547.

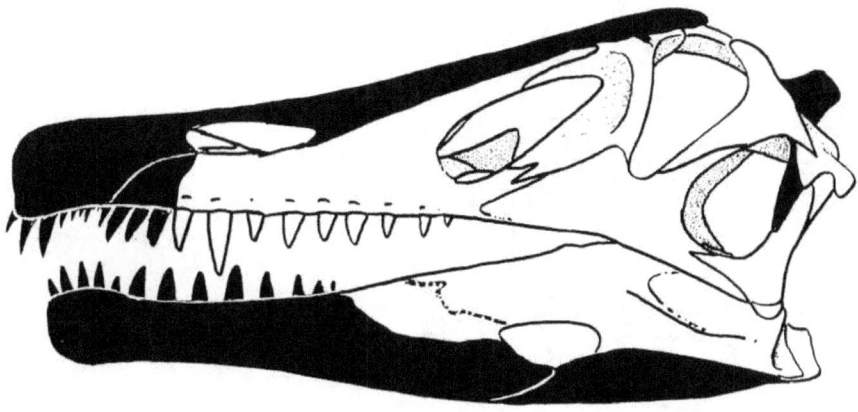

Figure 5). New interpertation of *Irritator,* after Sues, et al., 2002.

I've been asked to comment a bit more on the new Czerkas volume. This will be a brief look at the othernew genera from that volume. Two pterosaurs are described, one from Utah the other from China. The Utah pterosaur *Utahdactylus kateae* (Czerkas, S. A., and Mickleson, D. L., 2002, The first occurrence ofskeletal pterosaur remains in Utah: In: Feathered Dinosaurs and the origin of flight, edited by Czerkas, S. J., The Dinosaur Museum Journal, v. 1, p. 3-13.). This is a very fragmentary specimen that is still in the matrix. The bones aren't well preserved and it is impossible to extract the specimen from the stone (editors note, a year before Stephen Czerkas passed away he sent an abstract of a new paper describing *Utahdactylus* to a friend of mine to edit, which was taken out of the matrix). *Utahdactylus* comes from the lowest member of the Morrison Formation, the Tidwell Member. This is the first skeletal remains of a pterosaur from that age from Utah. It is believed to be a rhamphorhynchoid, but other than that nothing more can really be said for the specimen.

The other pterosaur is also a rhamphorhynchoid, but is better preserved. *Pterorhynchus wellnhoferi* (Czerkas, S. A., and Ji, Q., 2002, A new rhamphorhynchoid with a head crest and has complex integumentary structures: In: Feathered Dinosaurs and the origin of flight, edited by Czerkas, S. J., The Dinosaur Museum Journal, v. 1, p. 15-41.) This is a very interesting specimen that is nearly complete. It comes from the Daofugou Village, Chefeng County, Inner Mongolia, Haifangou Formation, possibly from the Middle to Late Jurassic. The skull has a large crest with the leading edge being keratinous (Figure 6). The crest also shows coloration pattern. The teeth are small. The specimen also shows some of the wing membrane as well as 'hair' like plumage found though out the specimen. The soft tissue can be seen in ultraviolet light. The authors argue that these 'hair' like plumage should be more considered feathers like the feathers found in the theropods previously discussed in this article. The 'feathers' show the same structure and some have a 'bulb' and has several 'hair feathers' sprouting from the bulb. They also argue that because the 'hair feathers' are like of similar structure than pterosaurs must have a similar ancestor as dinosaurs (though I disagree with this).

81

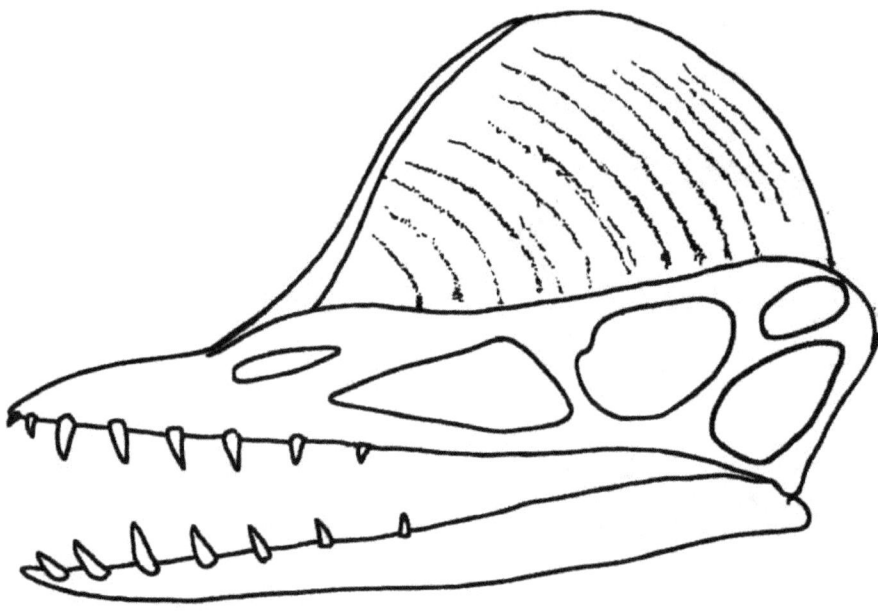

Figure 6). *Pterorhynchus wellnhoferi* skull showing the coloration pattern.

Two more birds were also described; *Archaeovolans* and *Omnivoropteryx. Archaeovolans repatriatus* is the front half of the '*Archeoraptor*' specimen (Figure 7). Even though Storrs Olson had described the tail as the type of *Archeoraptor* in an in house news letter for the Smithsonian, no one that I know of has used that paper. Also it has been demonstrated that the tail is part of *Microraptor* and the name *Archeoraptor* (if it is valid) would become a synonym of *Microraptor*. A deluded mess to be sure. The specimen comes from Xiasanjiazi (?), Chaoyang County, Liaoning Province, China, and is from the Jiufotang Formation, Early Cretaceous. The front half of the specimen is from a valid specimen and belongs to a basal ornithurine bird. The skull is long (badly crushed) and has 4 large premaxillary and 18 maxillary teeth, and an equal amount of dentary teeth. The wings are large and the shoulder is very similar to modern birds though the sternum is primitive. The legs are shorter than the wings and the tail and feet are missing so those skeletal elements are unknown (Czerkas, S. A., and Xu, X., 2002, A new toothed bird from China: In: Feathered Dinosaurs and the origin of flight, edited by Czerkas, S. J., The Dinosaur Museum Journal, v. 1, p. 43-61).

Figure 7). *Archaeovolans repatriatus* skeleton.

      The other bird was preliminary described and is still buried in the matrix. The authors used x-rays to view the specimen and it shall be fully prepared before a better description is done. *Omnivoropteryx sinosusaorum* is a very nice specimen (Figure 8). It comes from the Shangheshou, Chaoyang City, Liaoning Province, China, Upper Jiufutang Formation, Early Cretaceous. The skull looks uncannily like *Caudipteryx* (without the teeth). The wings are very large in comparison to the legs, the tail is very short and the foot has a retroverted toe, used for perching. It was a strong flyer. It is believed to have been a omnivorous (thus the name). There is a slight possibility that *Omnivoropteryx* was a flying form of oviraptorid, though this is speculation (Czerkas, S. A., and Ji, Q., 2002, A preliminary report on an omnivorous v6olant bird from northeast China: In: Feathered Dinosaurs and the origin of flight, edited by Czerkas, S. J., The Dinosaur Museum Journal, v. 1, p. 127-135).

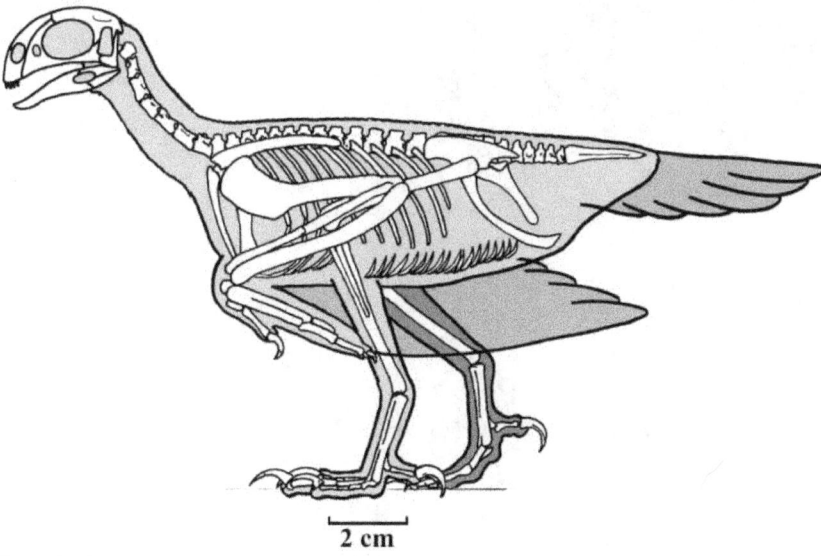

2 cm

Figure 8). *Omnivoropteryx sinosusaorum* skeleton.

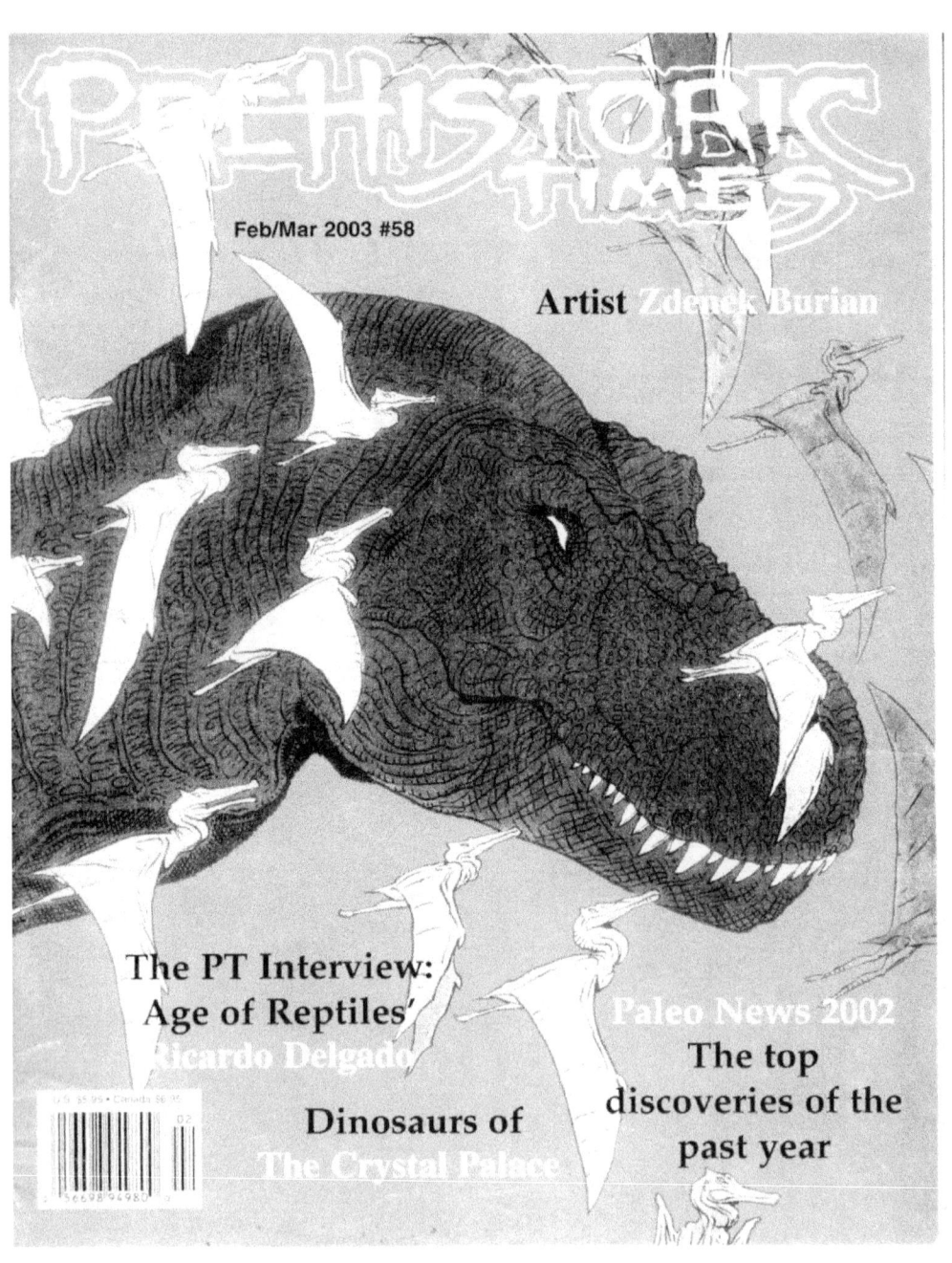

PREHISTORIC TIMES

Feb/Mar 2003 #58

Artist Zdenek Burian

The PT Interview:
Age of Reptiles'
Ricardo Delgado

Paleo News 2002
The top
discoveries of the
past year

Dinosaurs of
The Crystal Palace

Ford, T. L., 2003, How to Draw Dinosaurs. Arm waving in theropods (part one): Prehistoric Times, n. 58, p. 14-15.

Chapter 12

Arm waving in theropods (part one).

Recently Ken Carpenter wrote a very interesting article on the wrist and arm movements of theropods. I really liked this article and thought that it'd make a great article for PT.

This will be a two-part article because of the large amount of illustrations needed for definitions (this part) and his conclusions (which theropod group moved their arms and wrists which way) and because of use of skeletal illustrations (for the second part).

Ken used the following definitions for the movements of the manus and arms.

Arms.

Abduction: The movement of the forelimb away from the body.

Adduction: The movement of the forelimb towards the body (e.g., forearm raised and lowered as in flight).

Extension: Increasing the angle of the forelimb (e.g., bending and straightening the forearm at the elbow).

Flexion: Decreasing the angle between adjacent bones at a joint (e.g., from a bent or lowered hand to making the arm and hand into a straight line).

Pronation: Rotation of a horizontal or near horizontal forelimb so the palm faces downwards (ventral direction).

Protraction: Moving the forearms forwards (anterior direction) parallel to the midline of the body.

Retraction: Moving the forearms backwards (posterior directions).

Supination: Rotation of the forearm in an upwards direction (dorsal direction).

Wrist and manus.

Extension: Moving the manus from a downward position to a straight line with the arm.

Flexion: Moving the manus from a straight line to the midline, and bending the manus at the wrist downward.

Hyperextension: Bending the manus from the straight line backwards (like you're making a stop motion with your hand).

Radial abduction: Moving the manus from the straight line away from the midline.

Ulnar adduction: Moving the manus from the midline to a straight line with the arm.

Fingers and claws (The claws themselves can be moved independently from the fingers).

Extension: Opening the fingers.

Flexion: Bending the fingers.

Hyperextension: Bending of the fingers or claws upward. The fingers and claws can move upwards slightly.

85

Some of the terms can be used for the ankle and pes. The major difference is that the ankle could not rotate or twist. It was held stiff and ridged and moved fore ward and aft and not side to side. The ankle did hyperextend (move forward), flex and extend (move backward). The phalanges of the pes and ungual could also do the same.

These definitions can be used for any dinosaur or animal movement and you may want to keep a copy of this nearby. Ken stresses that not all theropods were able to move their arms in the ways described and will be explained in the next article.

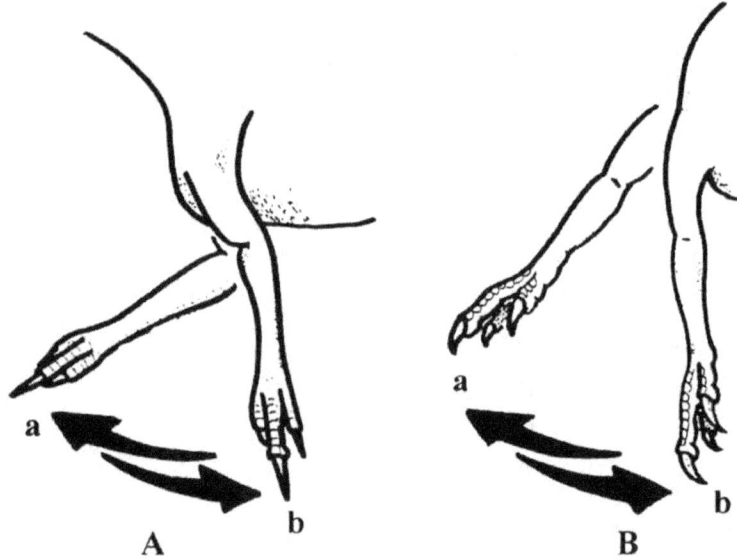

Figure 1). A). Flexion (a) and extension (b) (movement of the elbow); B). Abduction (a) and adduction (b) (movement of the shoulder) (after Carpenter, 2002).

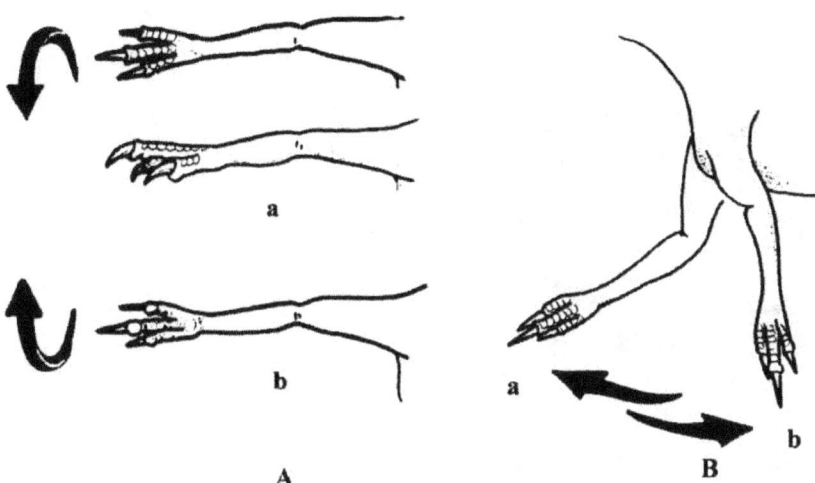

Figure 2). A. Pronation (a) and supenation (b) (movement at the elbow), B). Protraction (a) and retraction (b) (movement at shoulder) (after Carpenter, 2002).

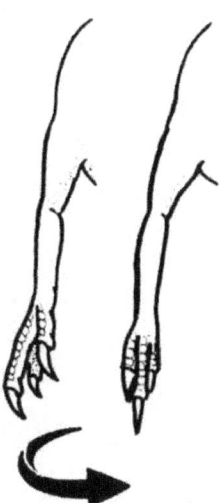

Figure 3). Rotation. (movement at the elbow) (after Carpenter, 2002).

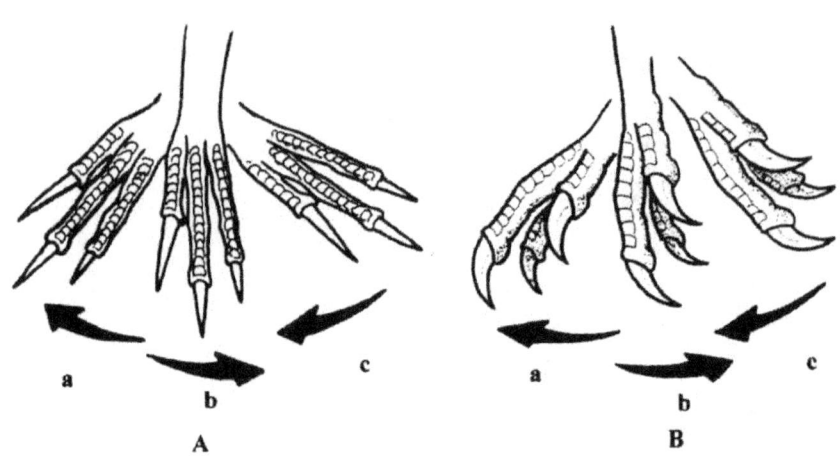

Figure 4). A). Radial abduction (a), flexion, (b) and ulnar adduction (c) (movement at wrist in plane of manus); B). Hyperextension (a), flexion (b) and extension (c) (lateral view of the movement of the manus) (after Carpenter, 2002).

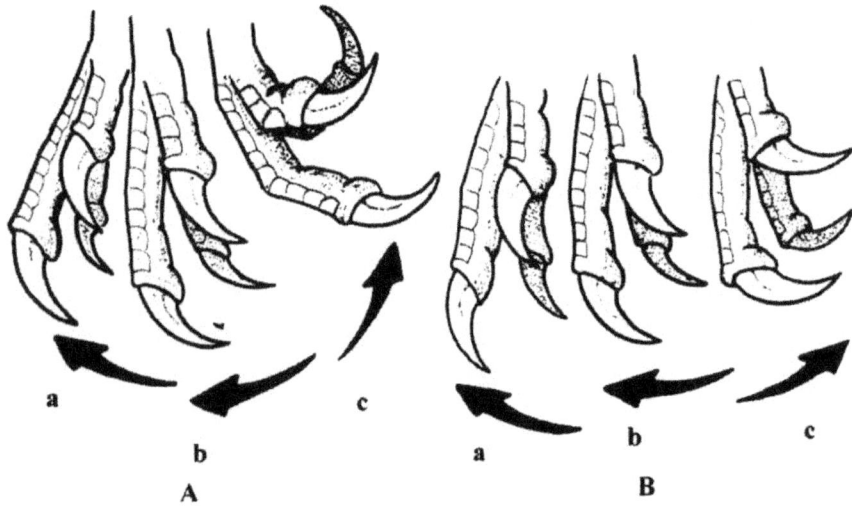

Figure 5). A). Hyperextension (a), flexion (b) and extension (c) of the digits (lateral view of the movement of the digits); D), Hyperextension (a), flexion (b) and extension (c) of the claws (lateral view of the movement of the claws) (after Carpenter, 2002).

Bibliography

Carpenter, K., 2002, Forelimb biomechanics of nonavian theropod dinosaurs in predation: In: Concepts of functional, engineering and constructional morphology, edited by Gudo, M., Gutmann, M., and Scholz, J., Senckenbergiana Lethaea, v. 82, n. 1, p. 59-76.

PREHISTORIC TIMES

Apr/May 2003 #59

The PT Interview:
Jurassic Park's
Mark "Crash" McCreery

PT Interview:

artist **Phil Wilson**

Znedek Burian Part 2

Canadian Dinosaur Trackways

Ancient Creatures of the
Burgess Shale

U.S. $5.95 • Canada $6.95

04

0 56698 94980 0

Ford, T. L., 2003, How to Draw Dinosaurs. Winged dromaeosaurs: Prehistoric Times, n. 59, p. 14.

Chapter 13

Winged dromaeosaurs

  As I mentioned in my PT article on feathered dinosaurs (PT 57) I'm reporting on a recent description of a small theropod with feathers on its legs. I alluded to this find in that article but was unable to comment on it until it was published. The new small-feathered theropods were described in a Nature article by Xu, et al. (2003). In it they describe a new species of *Microraptor, M. gui*. There are two nearly complete specimens (one is 77 cm long and has the most of the skull missing so the complete length of the skull is unknown) (Figure 1, 2). Both have uncinate process on the ribs like birds and some theropods. The top of the head had long pennaceous feathers (as has been seen in other feathered theropods and Liaoning birds) and were probably used for display similar to some modern birds (Figure 3a). The wing feathers have the primaries and secondaries as well as convert feathers. The tail has pennaceous feathers near the tip (starting at caudal 15) as seen in *Cryptovolans* and other feathered theropods (Figure 3b). What is astounding about these new theropods is that the legs are feathered. I'm not talking about 'hair' like feathers that cover the body of the Liaoning theropods and birds or the NMG 999 (which is the one with splayed out arms and legs that I reported on in an earlier PT article), but asymmetrical flight feathers! (Figure 3c). The feathers are on the back of the metatarsals, tibia and femur. They have the same pattern, as do the wing feathers. These leg feathers are not an isolated case, but have been found on 4 other specimens (*Microraptor* sp, and dromaeosaur incertae sedis).

  What does this mean? Did they flap their hind legs like wings? No. It isn't know if they could even splay them out horizontally because the femoral heads aren't well preserved, but in the type specimen of *Sinornithosaurus* it's femoral heads are rounded which would allow greater movement of the hind legs. *Sinornithosaurus, Microraptor* and *Cryptovolans* all may belong to the same family (or even the same genus) so it is not unlikely all could splay out their legs and use them like wings or a gliding surface. (Figure 4). Xu, et. al. believed that these small dromaeosaurs were arboreal and climbed trees.

  They could have used their legs like rudders when they leapt out of a tree and flew/glided to the ground; using the legs in an inverted V in a similar fashion of some modern aircraft. The forewing feathers could not have overlapped the hind legs if the wings were used for flapping because the hind legs would have gotten in the way when they were held horizontally. If they just held their appendages (fore and aft) horizontally, the surface area of the wings would have been large and would be used in only gliding and not flapping. I personally don't think they glided but were active fliers.

  In *Cryptovolans* you can see the leg feathers coming off the thigh, and possibly of on the metatarsal. Greg Paul mentioned this to me and the Dinosaur Mailing List months ago. I reiterated what Stephen Czeraks told and showed me (I did see *Cryptovolans*) the feathers are from the wing and the leg is lying over it. Both Greg and Stephen are correct. The wing feathers can be seen and below that are the leg feathers, and I think (I'm not sure) I can see metatarsal feathers like those seen in *M. gui*. The problem is there isn't much 'meat' or attachment area for the feathers on the metatarsals so I'm not sure how the feathers attached. Greg Paul theories that the metatarsal feathers were movable because otherwise they would have gotten in the way when the animals walked on land or would be a hindrance in flight because they would not be in the optimum angle for flight. Another hypothesis is they may have been used in display (Figure 5).

  Could other Dromaeosaurs also have had feathers on their legs? Or wings? I believe more than likely they all did to some degree.

  Who wouldn'a thunk it? A four winged dinosaur/bird. Actually C. William Beebe did in 1915 (Figure 6). While studying squabs (hatchling birds) he notices that several had feathers on their hind legs. He called this the pelvic wing (and should probably be used when describing theropods). The thigh and femur has a patagium and feathers. He studied the white-winged dove, domestic pigeons, jacana, and great horned owl. Only the hatchlings had the pelvic wing and as the bird grew this was lost. His theory was that a four-winged animal was an indeterminate form from a tree-climbing animal to a flying animal, which he named the Tetrapteryx stage. In 1916 Krosby illustrated a 4-winged *Archaeopteryx* for his Popular Science article and in 1917 Berry followed Beebe and Krosby. They all believed that the pelvic wing was a gliding apparatus and could not be flapped. In 1927 Heilmann also used a tetrapteryx stage (his *Proavis*) in his theory of avian evolution, though they all had a short pelvic wing feathers that extended from the thigh to the pelvis but not the metatarsals. Heilmann discounted Beebe's research because he was unable to duplicate his findings. A tetrapteryx form since then has been either forgotten or ignored. The Berlin *Archaeopteryx* specimen does show evidence of feathers on both the front and back of its thigh's. This has been debated for over 100 years, and depending on whom you talk to, they either had them or they didn't.

It isn't known just how these leg feathers evolved or if they had a role in the evolution of flight. Tetra-winged theropods may just be a dead lineage, or animals that played a crucial role in the development of the appearance of birds today. Only more specimens will help determine this.

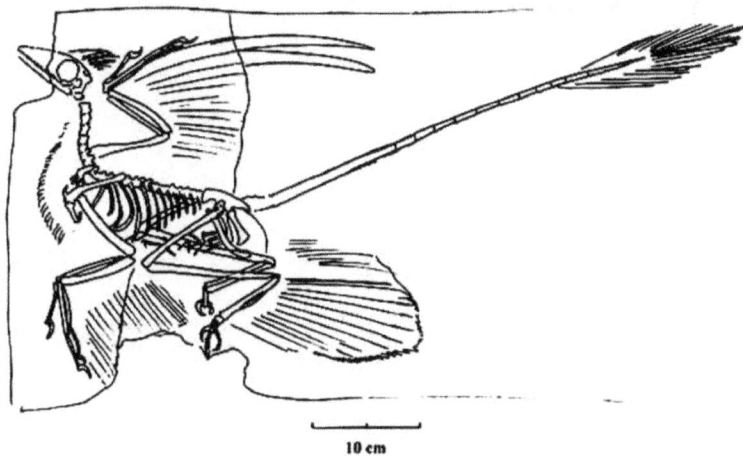

Figure 1). *Microraptor gui* in situ (after Xu, et al, 2003).

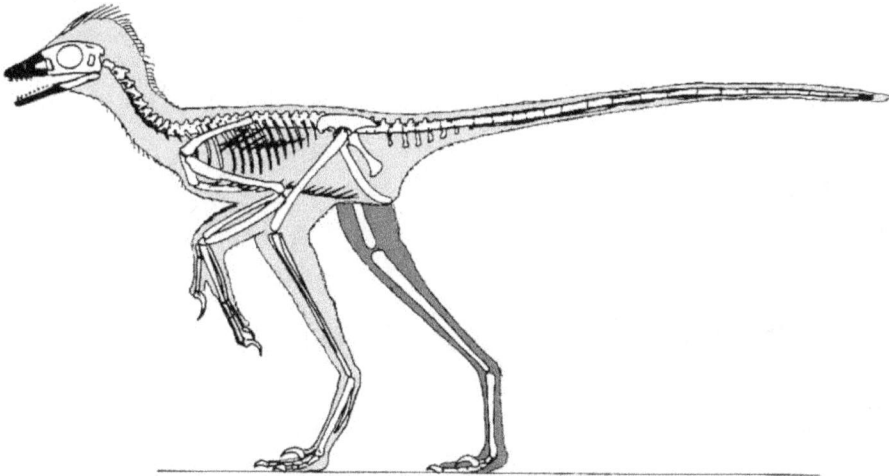

Figure 2). *Microraptor gui* skeletal reconstruction.

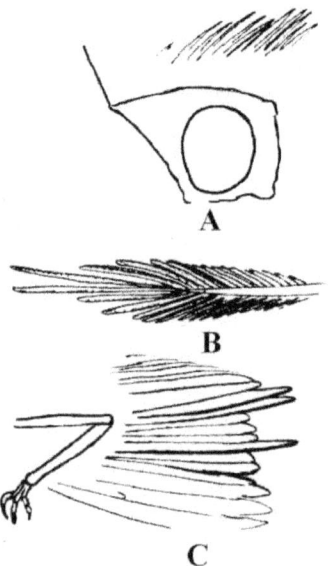

Figure 3). (A), skull showing feathers; (B) tail feathers; (C) feathers on the metatarsals (after Xu, et al, 2003).

Figure 4). Dorsal view of *Microraptor gui* (after Xu, et al, 2003).

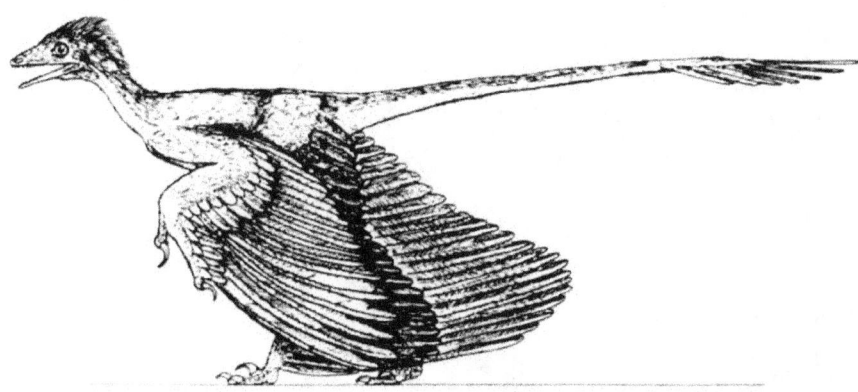

Figure 5). Life restoration of *Microraptor gui*.

Figure 6). Beebe's Tetrapteryx stage.

Bibliography

Beebe, C. W., 1915, A Tetrapteryx stage in the ancestry of Birds: Zoologica, v. 2, n. 2, pp. 38-52.
Xu, X., Zhou, Z., Wang, X., Kuang, X., Zhang, F., and Du, X., 2003, Four-winged dinosaurs from China: Nature, v. 421, p. 335-340.

PREHISTORIC TIMES

June/July 2003 #60

Big
Ten
Year
Anniversary
Issue

U.S. $5.95 • Canada $6.95

94

Ford, T. L., 2003, How to Draw Dinosaurs. Arm waving in theropods (part two): Prehistoric Times, n. 60, p. 14-15.

Chapter 14

Arm waving in theropods (part two).

Two issues ago I started this series of articles to explain how some theropods moved their arms. This stems from a recent article written by Ken Carpenter, the head paleontologist at the Denver Museum of Natural History. In the first article I went over the definitions and this one will cover the following theropods in his article; *Coelophysis*, *Coelurus*, *Allosaurus*, *Deinonychus*, and *Tyrannosaurus*. The easiest way to do this is to have the text also the figure captions.

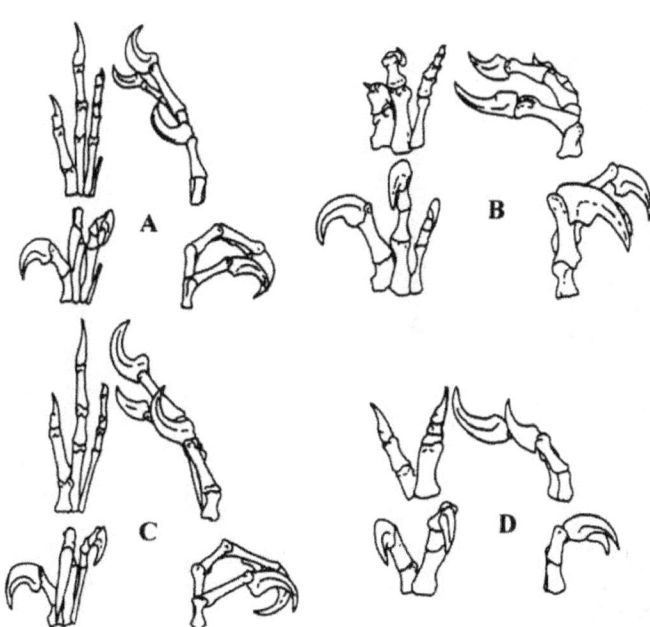

FIGURE 1): Starting at the manus; hyperextension and flexion (or movement of the fingers) in palmer (palm) and profile (lateral view); *Coelophysis* has the least amount of hyperextension and more flexion (A); *Allosaurus* has a greater amount of hyperextension and lest flexion (perhaps to use as a spring action into it's prey) (B); *Deinonychus* has little hyperextension but greater flexion (C); and *Tyrannosaurus* is more like *Allosaurus* (D) (after Carpenter, 2002).

95

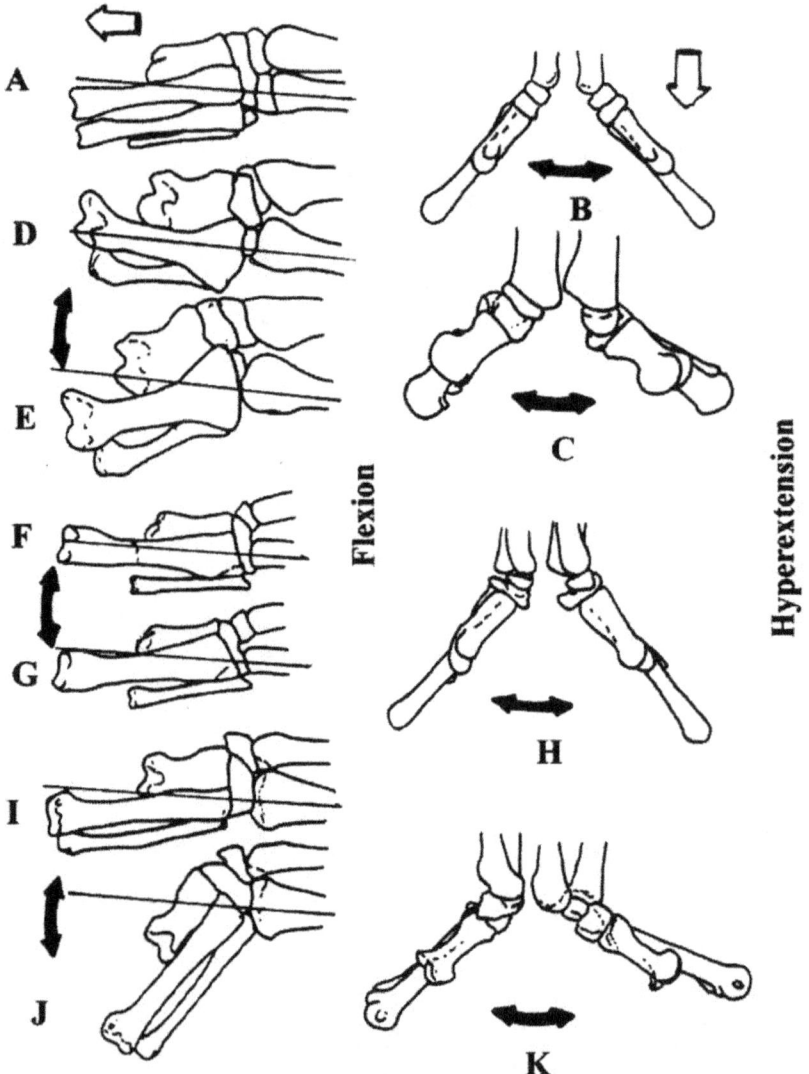

FIGURE 2): Moving up to the wrists; radial abduction-ulnar adduction and flexion-hyperextension. Theropods that lack a semi-lunate carpal had the least amount of movement (Figure 2). *Coelophysis* has the least amount of radial abduction-ulnar adduction (A) hyperextension and more flexion (B); *Allosaurus* has a greater amount of flexion-hyperextension 100 degrees (perhaps to use as a spring action into it's prey) (C), and about 20 degrees of ulnar adduction (D, E); The following two have semi-lunate carpals; cf. *Coelurus* has about 5 degrees radial abduction-ulnar adduction (F,G), and about 75 degrees of flexion-hyperextension, which are equal (H); *Deinonychus* has 50

degrees of radial abduction-ulnar adduction (I, J), and 115 degrees flexion-hyperextension in which 70 degrees is hyperextension (K) (after Carpenter, 2002).

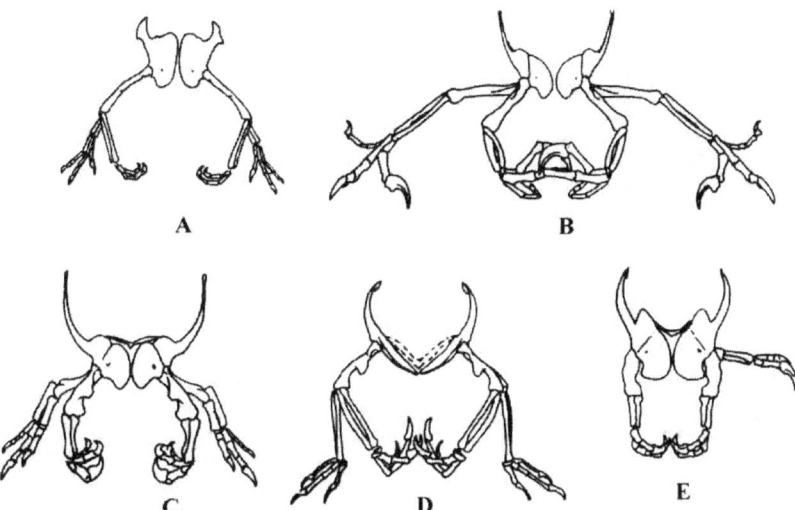

FIGURE 3): Forelimb movement in anterior view; maximum abduction with hands at hyperextension and at anteromedial adduction with hands flexed. *Coelophysis* has little movement (A); *Coelurus* has much greater movement and is able to bring the manus to the midline of the body (B); *Allosaurus* has a little less (C); *Deinonychus* can bring the manus to the mid line (D). Ken Carpenter does not believe dromaeosaurids couldn't move their arms like those of birds (i.e. the arm above the body). Greg Paul disagrees and believes they had the same amount of movement as birds. The ability to do this is the position of the glenoid (shoulder socket) and a rounded humeral head. Rear facing glenoid would not allow great movement, but a laterally placed glenoid would (as in modern birds). This is where Ken and Greg disagree, Ken believes the glenoid in dromaeosaurids wasn't lateral while Greg does (I'm more inclined to believe Greg in this); *Tyrannosaurus* has more lateral movement and bring the hand to the midline (E) (after Carpenter, 2002).

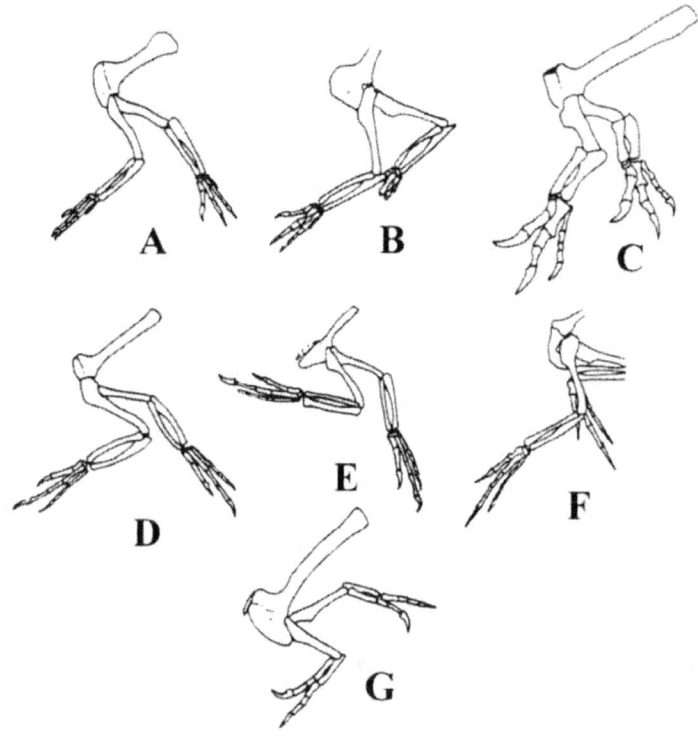

FIGURE 4): Forelimb movement in lateral view, protraction and extension of the shoulder, and maximum flexion and extension of the elbow; *Coelophysis* has little (A) though Greg Paul gives *Syntarsus* a bit more movement (B); *Allosaurus* has more (C); *Coelurus* has more (D); *Deinonychus* even more (E) while Greg Paul shows the humerus more vertical (E); *Tyrannosaurus* has more shoulder protraction (G) (after Carpenter, 2002).

Bibliography

Carpenter, K., 2002, Forelimb biomechanics of nonavian theropod dinosaurs in predation: In: Concepts of functional, engineering and constructional morphology, edited by Gudo, M., Gutmann, M., and Scholz, J., Senckenbergiana Lethaea, v. 82, n. 1, p. 59-76.

This issue's spotlight dinosaur is *Allosaurus*. I thought I'd add a few comments. The *Allosaurus* that just about everyone uses is from the Cleveland-Lloyd Quarry that Jim Madsen studied. His monograph has a complete skull illustrated. The problem is that that skull is a composite (also his cast that he sells) and the skull of Allosaurus looks a bit different. I've drawn that skull (A), the complete skull at Dinosaur National Monument (B), the complete skull from Howe Quarry (now at a museum in Europe) (C), and the skull from the Smithsonian (D). This specimen has a shorter higher skull with a pointed lachrymal horn. Because of this Greg Paul wanted to make it a new species. But there is a problem with the skull. After studying the mounted specimen at the Smithsonian I noticed that the skull was improperly assembled. There is a gap between the jugal and lachrymal and the skull should be shorter, or so I believed. When adjusting the size of the lachrymal to that of the How Quarry *Allosaurus* the skull is at the right height, but still to short length wise, it should be lengthened between the nasal, maxilla, and lachrymal, as well as the premaxilla. Yet, both sides of the skull show different things. The one side has the premaxilla, nasal, lachrymal attached (or so it seems), then it did have a short skull, but the other side is missing and when drawn in proportion to the Howe Quarry *Allosaurus* the skull's nearly match (E). This would give the skull a more 'normal' longer skull. Gilmore noted that the lower jaw belongs to a different specimen, and is longer than the skull (as mounted), but if the skull is lengthened then it be to short. I had thought the lower jaw of *Labrosaurus ferox* belonged to that skull, but I'm not so sure now. I'm not sure what to make of it now.

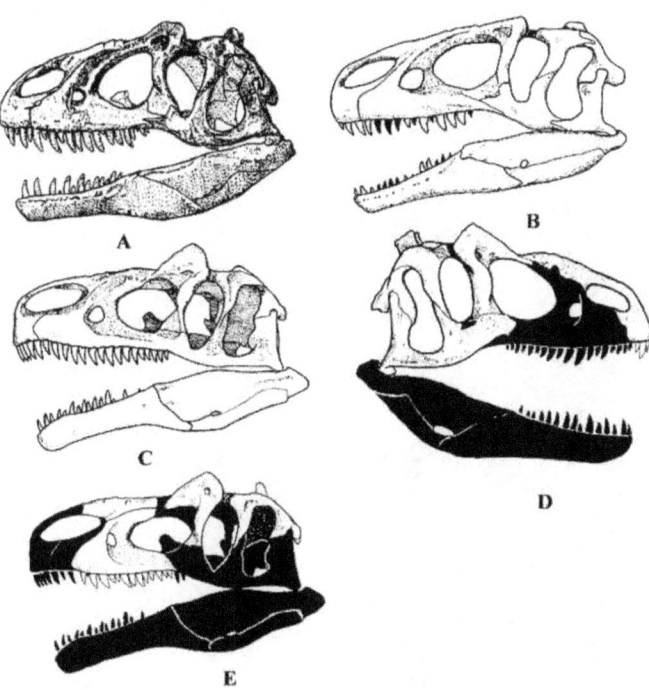

PREHISTORIC TIMES

Aug/Sep 2003 #61

PTEROSAURS!

PT Interview: Kevin Padian
TERROR BIRDS!
Cannibal Dinosaur!
Dino model
reviews & more!

U.S. $5.95 • Canada $6.95

100

Ford, T. L., 2003, How to Draw Dinosaurs. Pencil necked hadrosaurs?: Prehistoric Times, n. 61, p. 14.

Chapter 15

Pencil necked hadrosaurs?

Did hadrosaurs have a 'Pencil' neck like a swan, a horse neck, or a wrestler's neck? To answer this the first place to look isn't the back of the head but the front of the back. Long before the 'Dinosaur revolution' of the 1970's hadrosaurs both hadrosaurs and lambeosaurs were thought to have a straight back and a straight neck. Just look at the *Anatotitan* skeletons at the American Museum of Natural History (AMNH). One is standing on its hind legs, the other on all fours. The necks are straight. They were believed to have walked primarily on their hind legs and just feed on all four (not to mention the old aquatic lifestyle). But with the revolution they went from walking on just the hind legs to all fours. This changes how they held their necks dramatically because the front half of the body needs to be much lower to the ground than if it had a straight back. The articulated and partially mummified *Corythosaurus* on display at the AMNH shows this the best. The front of the back actually curves downward which brings the forelimbs closer to the ground and the neck curves up like a swan. This should be the normal position of hadrosaurs.

Now that we have that established, what about the neck? How did it look? Greg Paul gives hadrosaurs a thick neck. The skin starts at the back of the skull and continues to the front of the back. I have problems with this because this would hinder the downward movement of the neck greatly.

What about a Horse? The neck of a horse is thick but is skeletally different than a hadrosaurs. The neck is straight and does not 'dip down' like a hadrosaurs so the comparison doesn't work well as some would like. (Figure 1). The nuchal ligament attaches from the back of the skull to the front of the dorsal neural spines (or the withers). Muscle attach from the sides of the cervical vertebrae to the nuchal ligament (this gives the horse its think neck) (Figure 2). If hadrosaurs were built the same the nuchal ligament would have been shorter but the neck muscles would have been thicker and would be a hindrance when the animal lowered its head.

So should they have a 'Pencil' thin neck? Is there any support for that or any other look? Yes. There are some mummified specimens that show how the neck looked. The AMNH *Corthyosaurus* shows the neck curved and some skin starting at about 1/4 the length of the neck or just below the skull (Figure 3); also the neck is curved more than should due to rigor mortis. Also, Leonardo (the Mummified *Brachylophosaurus*), at the Malta Museum has skin from the middle of the neck. There is no evidence for the thick horse like neck or nuchal ligament for that matter. Steven Czerkas gave a talk at the 1993 SVP and came to the conclusion that they had horse-like neck and not a goose-neck. I think his terminology is incorrect, but his conclusion is. I believe hadrosaurs lacked the 'thick' neck Greg Paul gives them, but had a thicker neck than a pencil neck.

Also, there is a possibility hadrosaurs had a dewlap or pouch under the chin. At the 2002 Tucson Rock/Fossil show there was a new *Edmontosaurus* that showed this (and I had mentioned that before) so hadrosaurs had a really thick looking neck.

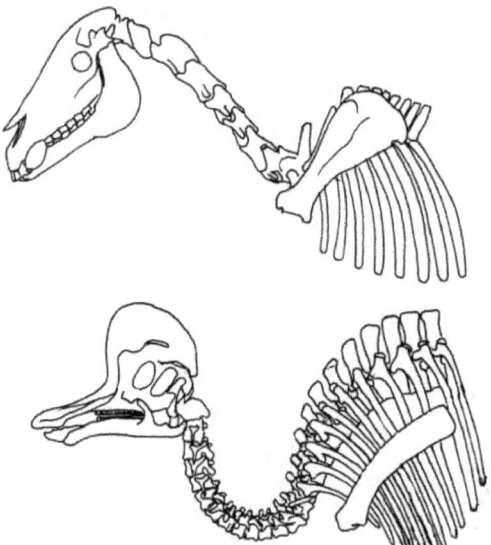

Figure 1). Top skeleton of a Horse, Bottom skeleton of *Corythosaurus*.

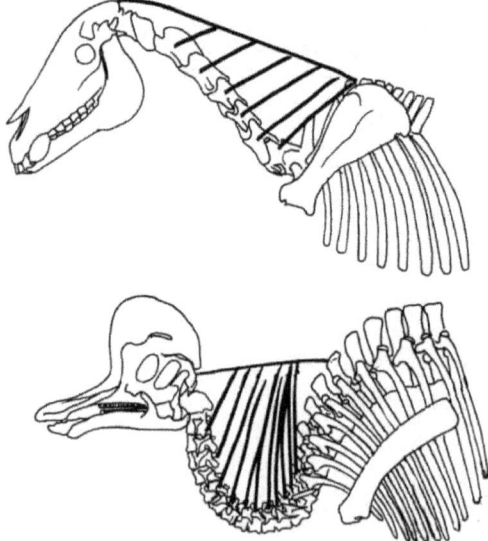

Figure 2). Top, diagram showing the tendons and muscles of the neck of a horse, bottom possible (though unlikely) the same for *Corythosaurus*.

102

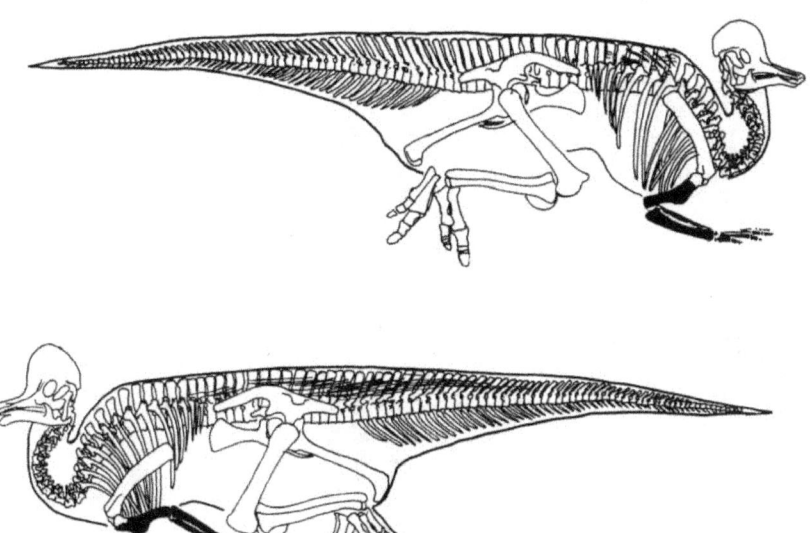

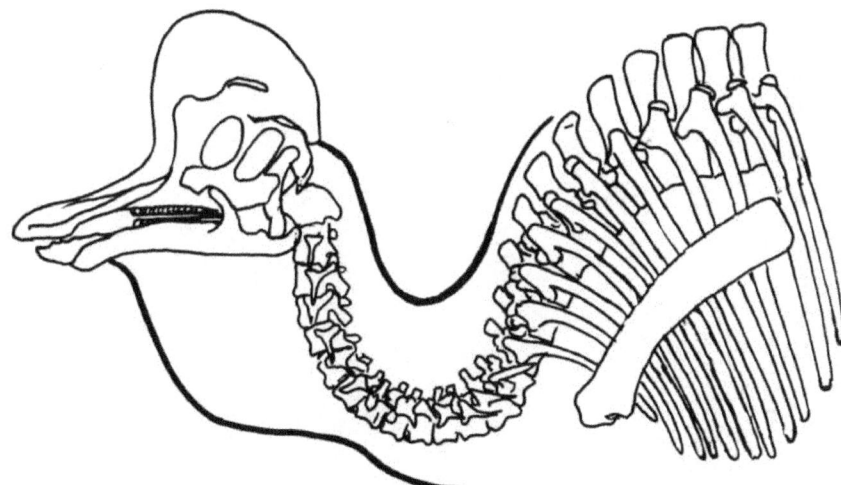

Figure

3). Side views of the mummified *Corythosaurus* at the American Museum of Natural History.

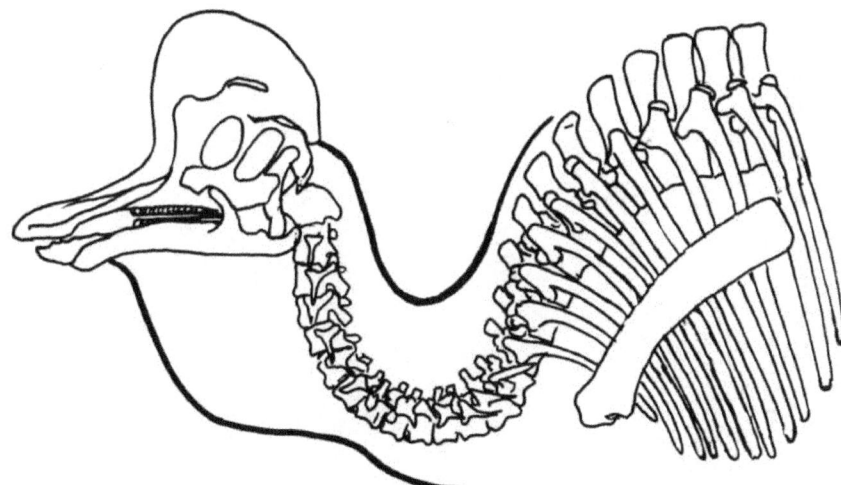

Figure 4). Side view of *Corythosaurus* showing possible dewlap and thickness of the neck.

103

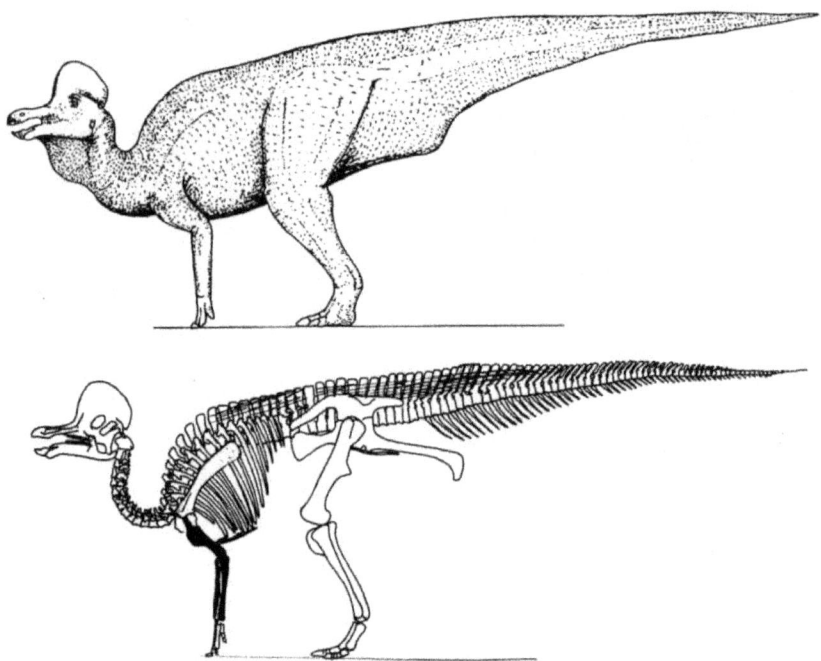

Figure
5). Top, Life restoration of *Corythosaurus* with the 'thicker' neck and a dewlap, Bottom, skeletal reconstruction.

Bibliography

Brown, B. B., 1916, *Corythosaurus casuarius*: skeleton, musculature and epidermis: Bulletin of the American Museum of Natural History, v. 35, p. 709-716.

PREHISTORIC TIMES

Oct/Nov 2003 #62

patagonian artist
Jorge Blanco

Gregory S Paul

Your favorite dinosaur books
& models as a kid!
Loch Ness Monster Movies!
The Disney Dinosaur Movie
that could have been!

105

Ford, T. L., 2003, How to Draw Dinosaurs. Claws: Prehistoric Times, n. 62, p. 14.

Chapter 16

Claws

Claws come in a variety of sizes and shapes for different functions; grabbing, cutting, climbing, walking, etc... Claws should be an easy thing to illustrate, sculpt, etc, though there are some things to keep in mind.

Claws have a groove on either one side or both and can be parallel or offset, and extend to the tip of the claw or be short of it (figure 1A, C). When I was young I thought this was a 'blood groove' (and others had expressed the same belief to me), i.e. when the claw was dug into the side of an animal it held the claw in that position and blood flowed from the animal. In actuality it's for a vein which supplies nourishment to the claw sheath. If you have a cat or a dog and have trimmed their claws you know that they have a vein in the claw and you have to be careful not to accidentally cut it.

The sheath of the claw would have extended to about 20-30% further than the 'boney' claw (figure 1B, D). The grabbing/cutting claws would have come to a sharp point and dinosaurs may have used trees to sharpen their claws similar to cats, though this is pure conjecture.

Foot claws are wider and are nearly straight. In ornithopods the claws are 'hoof-like' (Figure 2). As the claws are scrapped on the ground they would have flattened the sheath tips.

Jim Kirkland suggested to me that if the foot of some theropods didn't have large scutellate tarsus (foot scales) but single scales over each metatarsal because some theropods had 'loose' metatarsals (not fused, as in Tyrannosaurids, ornithomimid's etc). This is because the metatarsals would have moved during the foot falls and a single large scutate scale would have rubbed against each other over the metatarsals (figure). Those with fused metatarsals may have had larger scutellate tarsus.

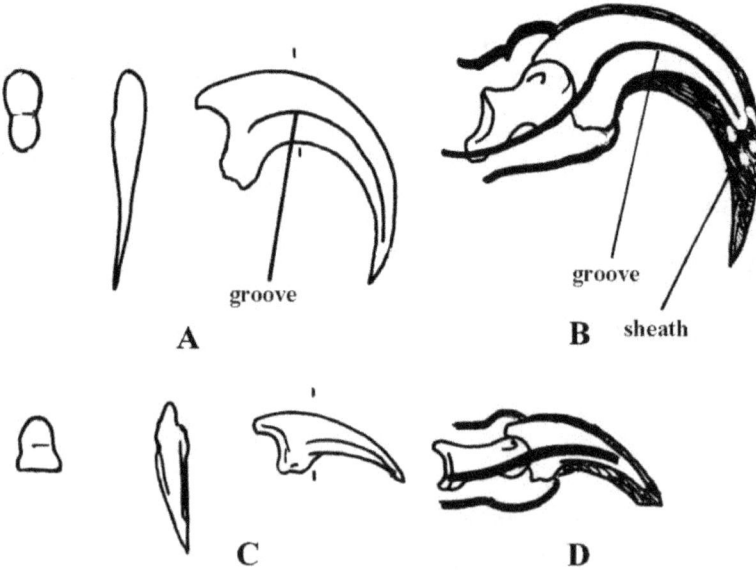

Figure 1. A) 'Killer'claw of *Deinonychus* in cut away view, top view and side view (groove marked, and hash lines are where the cut away view is taken); B) 'Killer' claw showing the vein and sheath; C) Toe claw of *Deinonychus* (same as A); D) Toe claw (same as B).

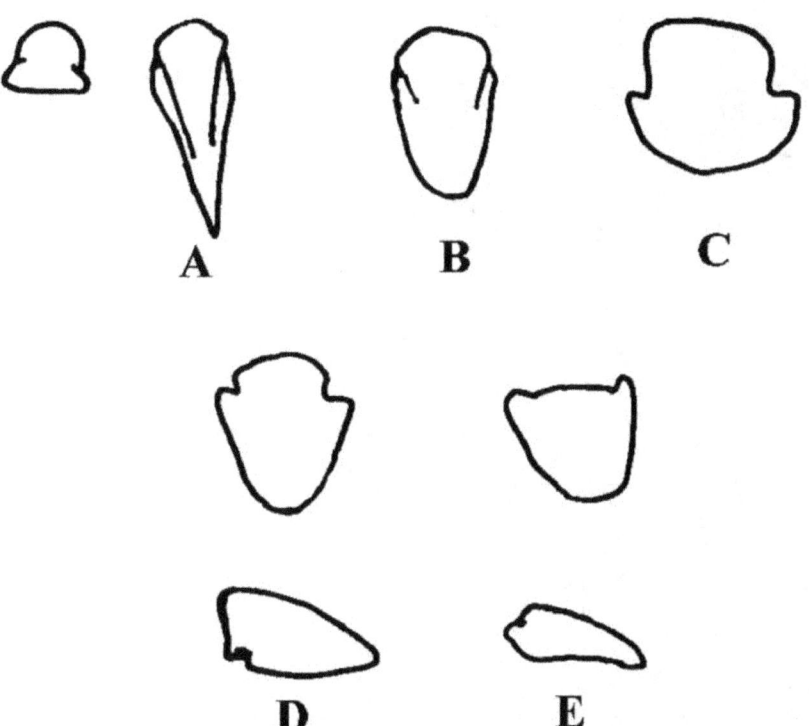

Figure 2. Comparison of toe claws (not drawn to scale); A) *Hypsilophodon*; B) *Iguanodon*; C) *Edmontosaurus*; D) *Centrosaurus* and E) *Stegosaurus*.

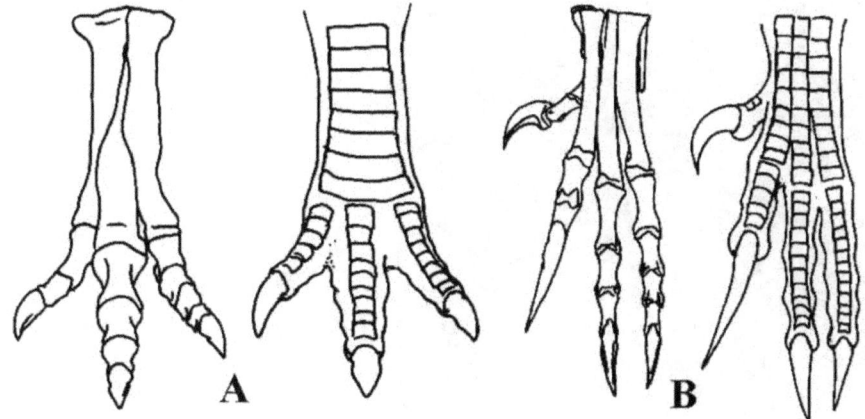

Figure 3. Metatarsals of showing possible scutellate tarsus. A) *Tyrannosaurus* metatarsals (tight metatarsals); B) Large scutellate tarsus; C) *Deinonychus* metatarsals (loose metatarsals); D) Individual scutellate tarsus.

GIANT PREHISTORIC CREATURES OF THE CARBONIFEROUS FOREST

PREHISTORIC TIMES

Dec/Jan 2004 #63

DON LESSEM ON THE
RISE & FALL OF THE
DINOSAUR SOCIETY

THE PT INTERVIEW:
DINO SCULPTOR
JUAN CARLOS ALONSO

FAVORITE
DINOSAUR MODELS
AS A KID, PART 2
& MUCH MORE!

U.S. $5.95 • Canada $6.95

Ford, T. L., 2003, How to Draw Dinosaurs. What is a feather? (part one): Prehistoric Times, n. 63, p. 18-19.

Chapter 17

What is a feather? (part one)

I was going to do this as one article, but the definitions ended up being so long that I've decided to make this a two-parter. Next issue we'll look at whether or not 'feathered' dinosaurs have the same type of feathers as modern birds. This issue will be a lesson in feather anatomy.

With all the new 'feathered' dinosaurs turning up I thought it'd be a good idea to look at just what feathers are, and the different types of feathers.

First, lets' look at the basic make up of feathers. I'm using Proctor & Lynch's book: Manual of Ornithology from Yale University Press (You can get it as soft or hard cover at Yale University Press's web site or Amazon.Com. I recommend this book). They have a great schematic of a flight feather (Figure 1).

A typical contour feather has a long central shaft and broad flexible vanes on either side of the shaft. The proximal central shaft is the calaus or quill. A mature feathers has a hollow tube (calamus) and the proximal tip of the feather is called the inferior umbilicus which is filled with pulp (I won't be going into the details of the inner soft tissue of feathers, unless asked to). Above the 'pulp' is the superior umbilicus, and above this is the hollow rachis. Major flight feathers have a square rachis. Flight feathers have asymmetrical vanes: i.e. the outside is short while the inside is long (editor's note: Only the first seven or seven eight feathers are asymmetric, the rest is symmetric). If you hold a feather and move the vanes they will separate into single 'lines' or barbs. The proximal end of the vane has plumaceous barbs; the distal vane is firm and bladelike and has pennaceous barbs. The barbs have a comma-shaped cross sections (Figure 1). Each barb is further divided into tinny branches called barbules. If you separate the barbs of a large flight feather you may be able to see smaller barbules though you'll probably need a high powered magnifying glass or microscope. The central shaft of each barbule is called the ramus. Each barbule has dozens of tiny projections called barbicels. These barbicels fasten together the surface of the feather vane. The distal barbules, the ones on the side that faces the feather tip each have 4 or 5 tiny hooklets or hamuli along their length. The hooklets of the distal-facing barbules catch the flanges on the proximal-facing barbules, similar to how Velcro works. Also each feather has a set of muscles in order to manipulate them not only in flight but in other daily activities.

There are about 9 different forms of feathers.

Contour feather (Figure 2a) cover the wings, including the large flight feathers (Remiges) of the wing and tail (Rectrices), and also covers the head, neck, body and legs. These feathers have barbules on the barbs.

Remiges (Figure 2b) are the flight feathers; primaries, secondaries, and tertiaries. Remiges are pennaceous contour feathers with prominent often asymmetrical vanes. Asymmetrical feathers help to indicate whether or not an animal could fly.

Rectrices (Figure 2c) are large, vaned, flight feathers on the tail.

Semiplumes (Figure 2d) are intermediated between the more pennaceous contour feathers and the strictly plumulaceous down feathers. They lie under the surface of contour feathers forming a smooth, aerodynamic body contours.

Adult down, or definitive down (Figure 2e) are extremely plumulaceous feathers and are an insulation underneath the contour feathers.

Natal down or neossoptiles (Figure 2f) feathers covers hatchling birds but are immediately pushed out by emerging juvenile plumage and appears as tufts at the tips of new feathers.

Powder feathers or powder down (Figure 2g) are special feathers with barbs that distinegrate into fine powder and are thought to aid in bird grooming and waterproofing its feathers. These feathers are never molted and grow continuously.

Bristles (Figure 2h) are contour feathers without vanes and are only a whiskery central rachis and tis nearly bare of barbs and barbules. These feathers are found near the eyes, the lores (between the mouth and the eye), nostrils and the rictus of the mouth. Not all birds have these type of feathers which are used mainly for protection.

Filoplumes (Figure 2i) are long, hair like feathers that monitor the position of the pennaceous feathers of the wings and tail. They are sensory feathers that help monitor during flight.

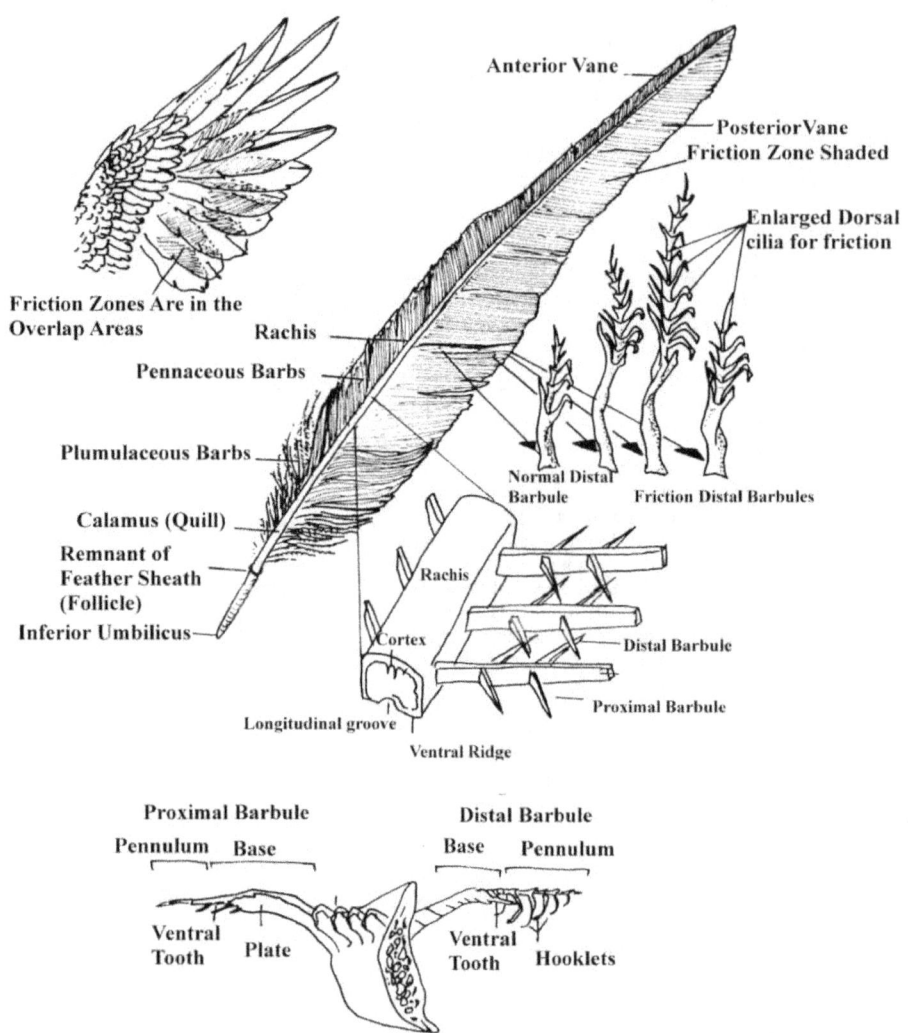

Friction Zones Are in the Overlap Areas

Anterior Vane

Posterior Vane Friction Zone Shaded

Enlarged Dorsal cilia for friction

Normal Distal Barbule

Friction Distal Barbules

Rachis

Pennaceous Barbs

Plumulaceous Barbs

Calamus (Quill)

Remnant of Feather Sheath (Follicle)

Inferior Umbilicus

Rachis

Cortex

Distal Barbule

Proximal Barbule

Longitudinal groove

Ventral Ridge

Proximal Barbule

Pennulum   Base

Distal Barbule

Base   Pennulum

Ventral Tooth   Plate

Ventral Tooth   Hooklets

Figure 1) Break down of a flight feather (after Proctor & Lynch, 1993).

110

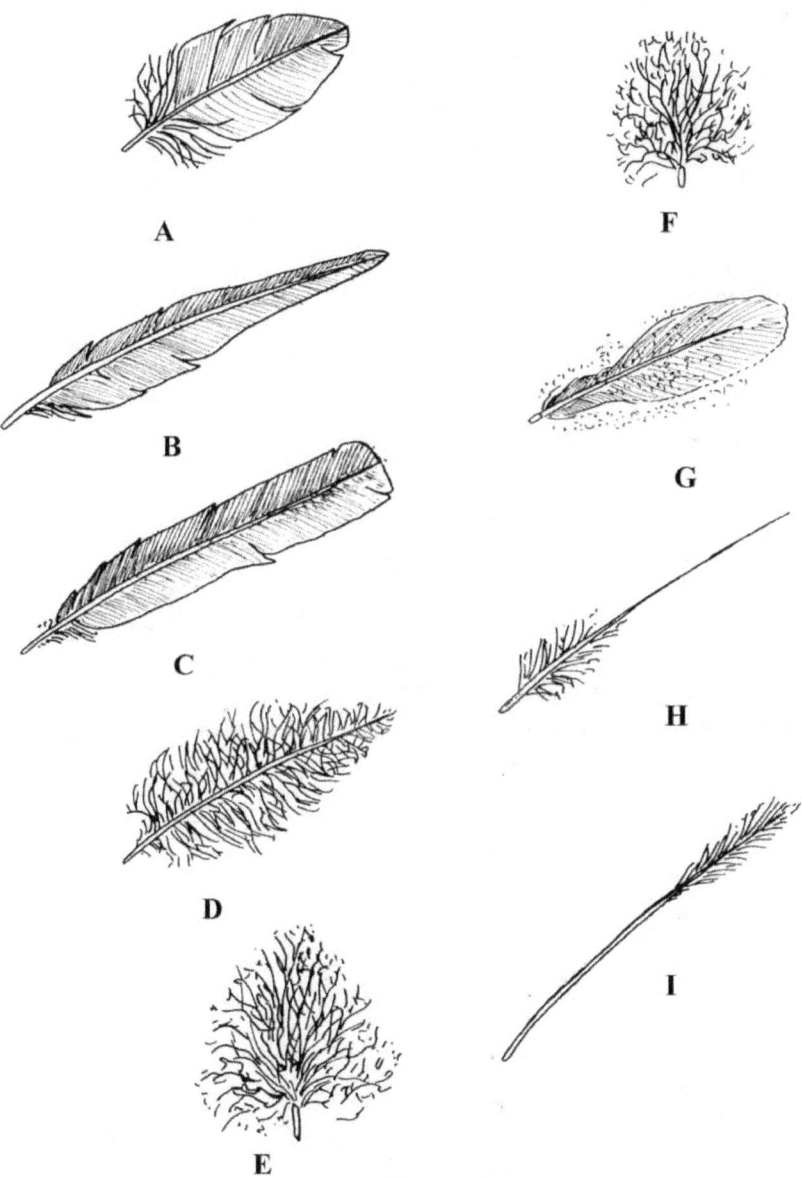

Figure 2a) Contour feather; 2b) Remiges: 2c) Rectrices; 2d) seimplumes. 2e) Adult down, or definitive down; 2f) Natal down or neossoptiles; 2g) Powder feathers or powder down; 2h) Bristles, 2i) Filoplumes (after Proctor & Lynch, 1993).

I won't go into how feathers develop (it'd take up too much space) I do recommend looking at it in Proctor & Lynch's books.

Prum has done extensive research on feathers, both evolution and function. His theory states that there are 5 possible stages of a feather evolution (Figure 3). I'll be using Prum veer Batum and then trying to explain each stage. I'm not sure I quite understand the definition Prum is using for feather growth. He calls the other proximal layer the follicle, which is hollow, but according to Proctor and Lynch, it's the sheath. Inside the sheath is the rachis, barbs, etc. Prum theorizes that the barbs did not evolve after the rachus, but before. The inner follicle is lined with barbules.

Stage 1, is the origin of the undifferentiated collar through a cylindrical epidermal invagination around the base of the feather papilla. This is just the follicle.

Stage II, Origin of the differentiation of the inner layer of the collar into longitudinal barb ridges and peripheral layer of the collar into the feather sheath. Just the barbs form the follicle (Sheath) and not rachis.

Stage III, either of these developmental novelties could have occurred first, but both are required before Stage IV.

Stage IIIA, origin of helical displacement of barb ridges and the new barb locus. Barbs forming from the rachis.

Stage IIIb, origin of paired barbules form peripheral barb plates within the barb ridges. Barbules forming off the barbs with no rachis.

Stage IIIa + b, origin of follicle capable of both helical displacement and barbule plate differentiation. Rachis, barbs, and barbules.

Stage IV, origin of differentiated distal and proximal barbules within barbule plates of barb ridges. Rachis, barbs, barbules and hooklets.

Commented [T1]:

Stage Va, origin of lateral displacement of the new barb ridges locus.

Stage Vb, origin of the division of posterior new barb locus into a pair of laterally displaced loci, and opposing anterior and posterior helical displacement of barb ridges toward the main feather and after feather of the follicle.
More next issue.

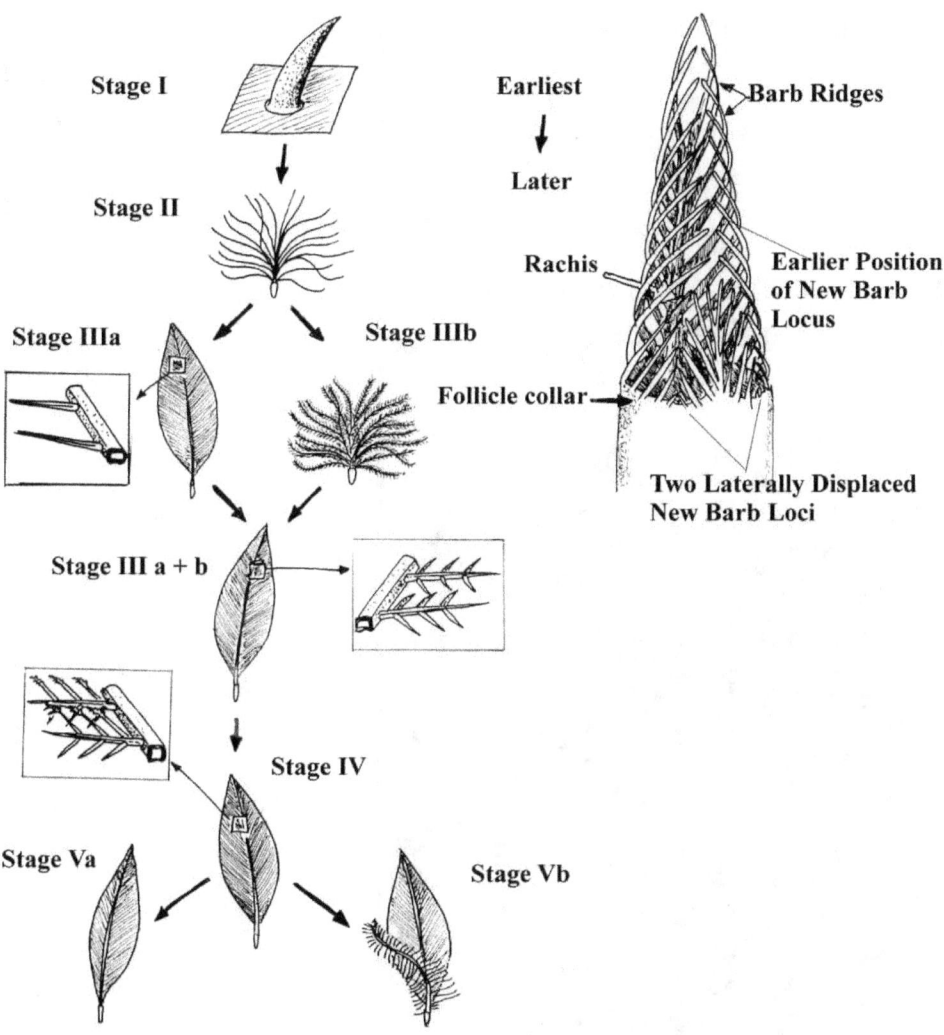

Stage I

Stage II

Stage IIIa

Stage IIIb

Stage III a + b

Stage IV

Stage Va

Stage Vb

Earliest

↓

Later

Rachis

Follicle collar

Barb Ridges

Earlier Position
of New Barb
Locus

Two Laterally Displaced
New Barb Loci

Figure 3) Evolution of feathers (after Prum, 1999).

Bibliography

Proctor, N. S., and Lynch, P. J., 1993, Manual of ornithology, avian structure & function. Yale University Press, 340pp.

Prum, P. O., 1999, Development and Evolutionary Origin of Feathers: Journal of Experimental Zoology (Mol. Dev. Evol), v. 285, p. 285-291.

La Brea Tar Pits and Sabertoothed Cats

PREHISTORIC TIMES

#64 Feb/Mar 2004

David Peters on pterosaur evolution
Fiction by Alan Dean Foster
Paleo News 2003 *The top discoveries of the year*

US $5.95 • Canada $6.95

Ford, T. L., 2004, How to Draw Dinosaurs. What is a feather? (part two): Prehistoric Times, n. 64, p. 18-19.

Chapter 18

What is a feather? (part two)

First I'd like to let ya'all know that I listed all my PT HTD articles and this one is #44. So, just a short year away I'll be writing my 50th article. Perhaps we can do a survey on the top ten with the readers?

I know I said this would be a two-part series but there is just too much important information that needs to be covered. I know I wrote a bit about the wing in PT 57 but I'll be going into more detail in this article.

For the most part this section I will be using a book by Jack B. Kochan; Birds; wings and tails (which actually belongs to a series of books, which I highly recommend).

When a bird flies with an outstretched wing the actual skeletal wing isn't straight but is bent. This is because of the patagium which is made up of the patagialis longus tendon that attaches the humeral head to the carpal bones along with several different muscles that attach to the tendon, and humerus (Figure 1). When this tendon is stretched out the skeletal wing is always bent. The pollex (thumb) is a single pointed bone (metacarpal 1 or II depending on who you follow, but I will be using I, II and III). An interesting note is that there are more modern birds known that have a claw on the wing (I'm not talking about the Hotzal) than is commonly known, but this claw is non-functional and usually covered up by the skin (the birds with this claw is a mutation). The skeletal wing hand is made up of metacarpal II that is usually fused to metacarpal III (forming the carpometacarpus). Mc III is bowed and is important when looking at the evolution of flight. After the carpometacarpus are the phalanges. The phalanges of mc II and III looks similar to the carpometacarpus. Finally the last phalange of digit 2 comes to a point.

The humerus can be short or long, depending on the bird. The radius is thin and straight and the ulna is bowed and most birds have 'knobs' on the back edge of the ulna where the secondary feathers attach and is a good indication whether or not an animal flew.

Feathers on birds don't for the most part grow randomly (except for penguins and toucans) but grow in tracts. The wing has the most named tracts (though not every ornithologist agrees on the number of tracts) (Figure 1). A feathered area is called the pteryla and featherless areas of skin between the tract are the apteria. The apteria is not entirely void of feathers. It shows signs of down feathers. Feathers within a tract grow in specific patterns; generally in overlapping rows and have a specific angle of orientation. Feathers within a tract remain in their relative position regardless of how the wing is moved. As the wing is opened the feathers will approximately remain in the same angle with respect to the segment they are attached to.

The top side of the wing is the scapulohumeral and the alar tract. The scapulohumeral tract has a narrow band of feathers on the shoulder (though some refer to two different tracts, the scapular and humeral. Since there is no apteria between them they appear as one tract). The scapulohumeral tract begins on the anterior or front edge of the scapula (pectoral region). It extends obliquely along the brachium (humerus) to the elbow (or following the body outline).

The next tract is the tertial which is on the posterior edge of the humeral segment (though some believe this to be part of the scapulohumeral tract). Some consider these feathers to be remiges. This tract can be short or long depending on the length of the humerus. The tertials and axillars grown nearly parallel to the trunk and the scapulars at a slightly outward angle and these feathers are symmetrical.

The rest of the top of the wing feathers are the alar tract, which consist of the alular quills, primaries, secondaries, and coverts. The coverts emanate nearly perpendicularly but have an extreme curvature so they appear to lay flat. The underside of the wing also contains the humeral and alar tracts. The humeral tract contains the axillary region (armpit area).

The flight feathers (the primaries and secondaries) grow only at the edge of the feather tract, while the converts grow over the entire tract (Figure 2). All visible feathers are considered contour feathers (which include the primaries and secondaries which are called collectively remiges). The primaries arise on the manus and point at an angle toward the outer edge. In many species the outermost primary is greatly reduced and is called the remicle or little remix. The primaries that are attached to the metacarpal are the metacarpal primaries. The remaining primaries are the digital primaries because they attach to the digits. The number of metacarpal primaries helps indicate which family birds belong to. Flamingos, grebes and storks have seven while nearly all others have six. The primaries of the hand are collectively called the pinion. It is interesting to note not all the wing feathers are asymmetrical, in fact only about first seven or eight are the rest are symmetrical.

The secondaries grow on the ulna which is the middle segment or the antebrachium/forearm which are technically called the cubitals which is Latin for elbow (Figure 2). The feathers on this tract grow slightly inward.

In a few species, there is an additional, smaller secondary feathers located in the space between the primaries and secondaries. According to Kochan this carpal remix seems to be disappearing through the evolution of birds and is still present in some gallinaceous birds and gulls. This may be important to the evolution of flight and early birds may have had these feathers.

The pollex has feathers on it known as the alula and is important when looking at the evolution of flight. Birds like *Archaeopteryx* which has a functional mc 1 and claw didn't have alula feathers. The alula feathers are slightly moveable and is used in flight. They vary in count from two in hummingbirds to as many as five or six in certain cuckoos.

How does a bird fold its wings without messing up the feathers? When the bird folds its wings the feathers naturally stack on top of one another and become compact next to the body (Figure 3), the primaries are stacked and are tucked under the secondaries. The secondaries are stacked with the innermost secondaries tucked under the scaplars and tertials. The stacking of feathers and associated muscle mass creates a prominent bulge at the shoulder and the scapular area. As Kochan notes (and this is where I've taken this information from) this mass should be a prime consideration when carving or illustrating any wildfowl, or dinosaur for that matter.

Next issue I WILL talk about dinosaurs and early birds.

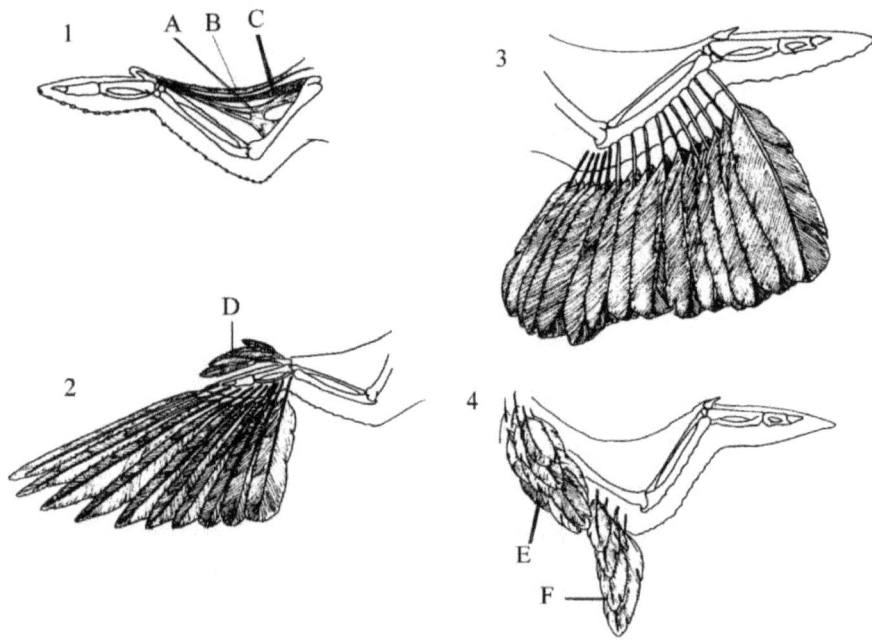

Figure 1). Wing and wing feathers; 1) Skeletal view of a wing; 2) Primaries; 3) Secondaries; 4) Humeral and scapular feathers; A) Biceps slip; B) Patagium; C) Patagialis longus tendon; D) Alular feathers; E) Scapulars; F) Tertials (after Kochan, 1996).

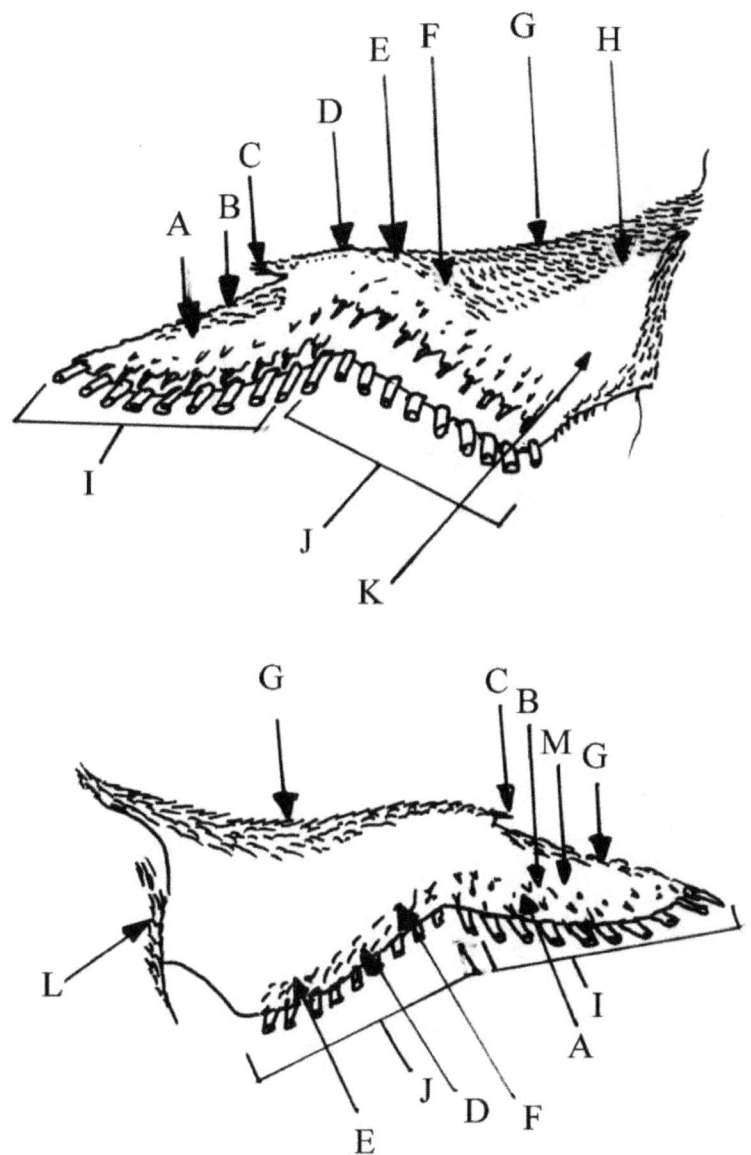

Figure 2). Top and bottom view of a wing showing feather tracts; A). Greater primary coverts; B) median primary coverts; C) Alular quills; D) Greater secondary coverts; E) Median secondary coverts; F) Lesser secondary coverts; G) Marginal coverts; H) Scapulohumeral tract; I) Primaries; J) Secondaries; K) Tertial region; L) Axillar region; M) Lesser primary coverts (after Kochan, 1996).

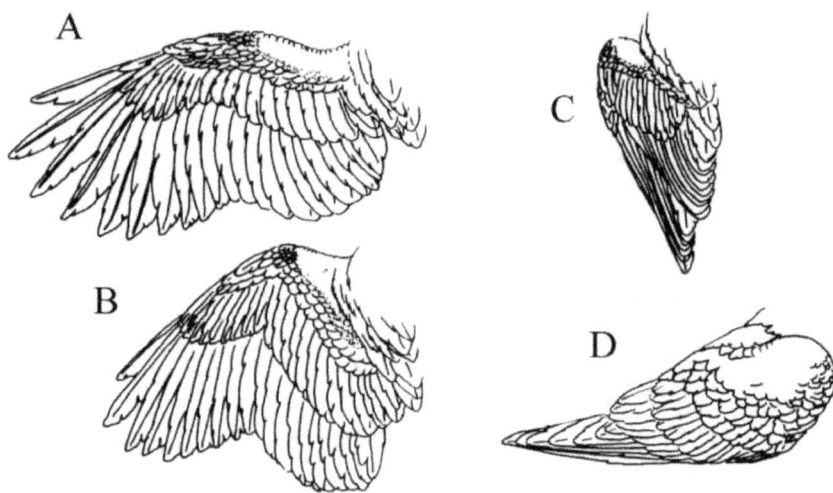

Figure 3). Folding of a wing. A) Out stretched wing; B) Partially folded wing; C) Folded wing top view; D) Side view of wing (after Kochan, 1996).

Bibliography

Kochan, J. B., 1996, Brids; Wings & Tails, Stackpole Books, 88pp.

Dino movies of Wah Chang by Mark Berry

PREHISTORIC TIMES

#65 Apr/May 2004

The PT Interview:
Sculptor
Bruce
Bowman

the eating habits of
Tyrannosaurus
rex

Paleontologist
Mark Norell

David Peter's
Pterosaurs part II

Make your own
Bambiraptor

U.S. $5.95 • Canada $6.95

04

0 56698 94980 0

119

Ford, T. L., 2004, How to Draw Dinosaurs. What is a feather? (part Three, conclusion), Theropod feathers: Prehistoric Times, n. 65, p. 18-19.

Chapter 19

What is a feather? (part three) Theropod feathers

The body (proto feathers)

Theropod feathers range from fluffy 'proto feathers' that cover the body to asymmetrical flight feathers. The first theropod found with feathers is *Sinosauropteryx*. I remember at the SVP (1995, in Pittsburgh) Mike Skrepnick came up to George Olshevsky and I and asked if we wanted to see pictures of the new feathered dinosaur. How could we refuse? That was the first time I saw *Sinosauropteryx*. To say *Sinosauropteryx* started a controversy is an understatement. It has been argued that the 'fluffy filament' structures surrounding the body are 'proto feathers', collagen fibers, musculoskeletal structures or fungal mats.

The main problem with the 'fluffy filament' structures is that they are mainly stains (carbonized filaments) on the slabs and not three dimensional impressions. Feathers are made up of beta keratin and it is impossible to check to see what kind of keratin the Liaoning theropods (or birds) had because they are stains (as Mary Schweitzer told me). I remember at one of the SVP Phil Currie came up to George Olshevsky and I and told us about a one-clawed theropod. He smiled and left us in the dark. It was *Mononykus* and Mary Schweitzer et al have studied *Shuvuuia*, a *Mononykus* relative (alvarezsaurid) and have found out that *Shuvuuia* had 'proto feathers' made up of beta keratin same as modern birds. I believe alvarezsaurid are theropods and not birds. Because these 'proto feathers' are similar in morphology to the Liaoning theropods/birds it can be surmised that the Liaoning theropods/birds 'proto feathers' are made up of beta keratin and similar to modern birds.

I believed the filament structures in the Liaoning theropods/birds are 'proto feathers' and will continue to call them as such. The 'proto feathers' can range from a few centimeters to more than 10 cm in length. They can be straight, curved or bunched up. At the new Feathered Dinosaur exhibit there is an undescribed feathered dromaeosaur that has long 'proto feathers' on its neck that makes it look like a fluffy collar. Whether these 'proto feathers' laid flat against the body in life or if the 'proto feathers' stood perpendicular to the body like a porcupine, or if the 'proto feathers' could be lower or raised, or because of post mortem the 'proto feathers' laid perpendicular to the body and are an artifact of preservation, is not known.

So where do the 'proto feathers' fall on the 'Prum' scale? Prum believes they are stage I or bristles, though they maybe also be stage II. The body of *Sinosauropteryx* has long 'proto feathers' without any apparent signs of belonging to regional tracks (as do other Liaoning theropods and birds). It isn't known if they were hollow, or if they had barbs because of the poor preservation. These 'proto feathers' are different from the body covering of 'modern' birds. It is a mystery why all the Liaoning theropods and birds have these kinds of feathers.

It is interesting that a lot of attention has been given to the theropods with feathers and very little on the Liaoning birds which seem to have the same stage I/II body feathers, though they are said to have had down feathers.

Sinosauropteryx. *Beipiaosaurus*, *Protarchaeopteryx*, *Scansoriopteryx*, *Shuvuuia*, *Caudipteryx*, *Cryptovolans*, *Microraptor*, *Sinornithosaurus* all have the 'proto feather' body coverings and only the last four have wing and tail feathers.

The oldest theropod known to have had 'proto feathers' is the Middle Jurassic *Scansoriopteryx* (not from the Early Cretaceous as orignally reported). This implies that 'proto feathers' had a long evolutionary history.

The wings (primaries)...

No theropod had a patagium; dromaeosaurids, troodontids, oviraptorids etc. (figure 1). In fact, the first birds didn't either; *Archaeopteryx*, confuciusornithoformes, and most enantiornithes. The birds that lack claws on the wing may be the ones that did have the patagium. A lack of a patagium reduces the wing area for lift (editor's note, *Archaeopteryx*, and other early avian theropods have now been found with a patagium, and possibly some theropods).

What do the wing feathers look like? For the most part they are symmetrical with a rachis and barbs like primary wing feathers of modern birds. (Figure 2). A few theropods (*Microraptor* and *Cryptovolans*) have asymmetrical proximal wing feathers and as noted before these types of feathers are an indication of flight. I do believe they could fly and climb. It has been argued that because the theropods lacked a retroverted hallux (except *Scansoriopteryx*) they could not climb, but if you can grasp with your fore limbs you don't need a reversed hallux. I'll leave that for a future article. *Scansoriopteryx* just has long 'proto feathers' on the wing.

Sinosauropteryx specimens lack any 'proto feathers' on the arms. It is more than likely they did have the arms and body completely covered in 'proto feathers'.

Even though *Protarchaeopteryx* lacks 'proto feathers', or for that matter any feathers on it's wings is not to say it never did, and it could very well have had them. It does have them on the tip of it's tail.

Beipiaosaurus is known from a fragmentary specimen (hands and distal portions of the ulna and radius are important to this discussion), have large patches of long 'proto feathers' that are arranged in parallel rows and almost perpendicular to the wing bones as does *Sinornithosaurus* (type and the one nicknamed 'Dave'). Some of these feathers have branching distal ends and appear to be hollow (stage I). *Sinornithosaurus* has multiple filaments that are joined together and has two types of branching structure. One is similar to avian downy feathers the other avian pennaceous feathers that lack identifiable barbules. Even though 'Dave' lacks flight feathers it is argued by Czerkas and Paul that those feathers just weren't preserved (or plucked) and *Sinornithosaurus* would have looked like *Cryptovolans*, and *Microraptor*.

The wing of *Caudipteryx* has at least 14 symmetrical primary feathers. It is not known if they had converts due to the preservation, but I believe any dinosaur with primary wing feathers also had converts. No scapulars or humeral feathers were found either and may not have had them. *Caudipteryx* had small wings and wing feathers and could not fly.

Microraptor gui has several feathers preserved; 'proto feathers' and primaries. The wing has asymmetrical distal primaires, and the rest of the primaires are symmetrical and not well preserved enough to determine there excat lenght (though the orginal authors have the wing feathers way to short IMHO). Secondaries are visable but coverts are not.

Cryptovolans may or may not be the same as *Microraptor* (their synonym is being researched) and has a very similar wing, leg, and tail feather morphology.

Archaeopteryx (I'm using the Berlin specimen) has primaries and covert feathers on its wings and some possible body feathers.

It is interesting to note that the wing claws in a lot of the theropods, birds and even pterosaurs have them facing toward the body and not away from it. I don't know the significance of that but have drawn the claws the way they were preserved. I've even seen specimens that have claws going both ways.

What about other dinosaurs that are in the families compsognathids, dromaeosaurids, troodontids and oviraptorids? Though some of the Liaoning specimens belong to those families no where else have feathers been found associated with skeletons. I use to fight against feathered theropods as a whole but due to recent discovers in China and Mongolia I had to rethink that. I know believe that yes, it is very likely if extremely probable that those families did have feathers of some sort; i.e. 'proto feathers' on the body and wing feathers. Whether they were short or long I don't know. These theropods may be secondarily flightless theropods and if so I doubt they had real wing feathers. Something else to think about, since these theropods have true avian feathers on their arms, should we refer to them as wings and not arms?

The legs (primaries?)...

Only a few theropods have any feather impressions on their legs (figure 3). Archaeopteryx has a few on the front of one of its leg. 'Dave' has long 'proto feathers' similar to those on the wing. *Cryptovolans*, *Microraptor*, the new undescribed theropod and a few I saw in Tucson this year all have leg and metatarsal feathers. These are similar to the symmetrical wing primary feathers. *Microraptor gui* has clear feather impressions on the back of its legs and metatarsals. The type *Cryptovolans* has feather impression on the upper and lower leg with some very light feather impressions on the metatarsals. The referred specimen clearly shows long metatarsal symmetrical feathers as well as from the rest of the leg. These feathers were originally thought to be from the wing and were misplaced during preservation. Now that it is known that they had metatarsal feathers they are more easily recognized in specimens. These leg feathers are longer than the ones illustrated in the paper on *Microraptor gui*. Could other dromaeosaurs have had leg feathers? It is possible. I believe the smaller genera had a higher possibility than their larger relatives.

The tail (rectrices)...

The tail feathers (rectrices) in theropods differ in shape and length (figure 4). Sinosauropteryx has only 'proto feathers' and has no indication of any other type of feather. Caudipteryx has a 'V' shape array of feathers, though this

121

may be from preservational bias and may have been a short fan. Microraptor and Cryptovolans has a long tail with long tail feathers on the tip but non on the sides of the tail. Archaeopteryx has a shorter tail to body ratio and has a tail fan along the length of the tail.

In conclusion, some theropods have primary wing feathers (some asymmetrical), tail rectrices, and stage I/II (from Prum) 'proto feathers' covering their bodies. Some could climb and some could truly fly. Larger theropods from the same families may have been secondarily flightless and had smaller wings, leg and tail feathers (if they had them at all).

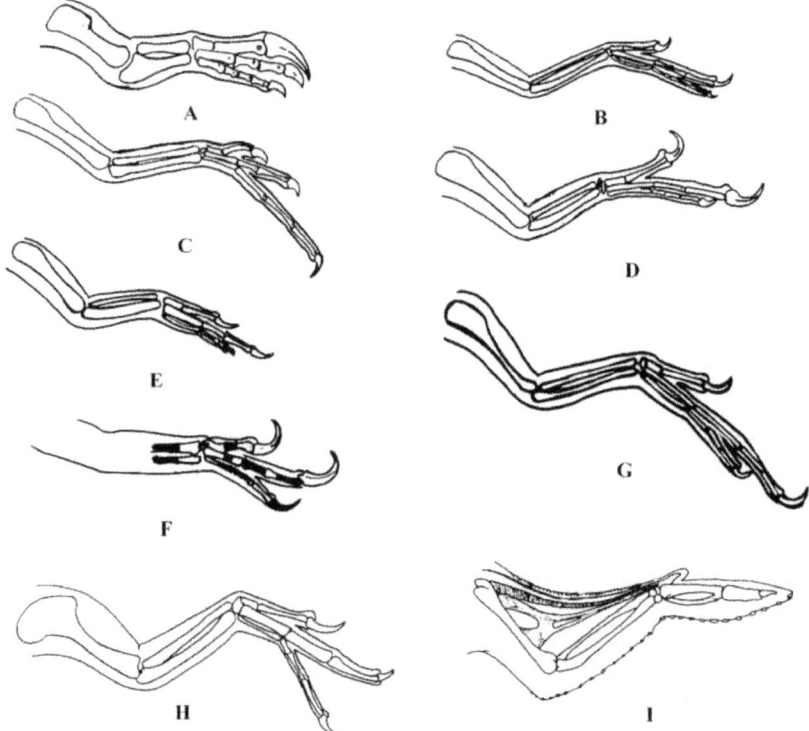

Figure 1). Arms of theropods showing no patagium; A) *Sinosauropteryx* (after Ji, & Ji, 1996); B) *Cryptovolans* (modified from Czerkas, et al., 2002); C) *Scansoriopteryx* (after Czerkas, & Yuan, 2002); D) *Protarchaeopteryx* (after Ji, & Ji, 1997); E) *Caudipteryx* (modified from Ji, et al., 1998); F) *Beipiaosaurus* (modified from Xu, et al., 1999), G) *Archaeopteryx* (modified from de Beer, 1954); H) *Confuciusornis* (modified from Chiappe, et al., 1999); I) modern bird (after Proctor & Lynch, 1993). (not drawn to the same scale).

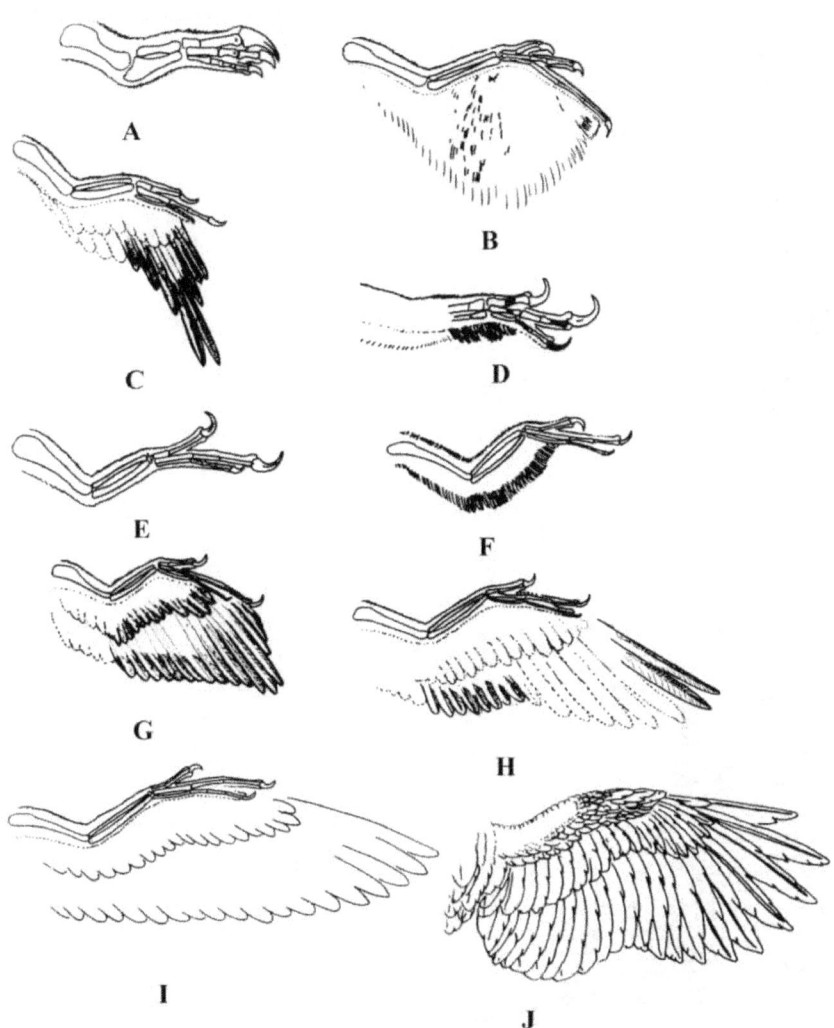

Figure 2). Wing feathers of theropods; A) *Sinosauropteryx* (after Ji, & Ji, 1996); B) *Scansoriopteryx* (after Czerkas, & Yuan, 2002); C) *Caudipteryx*(modified from Ji, et al., 1998)*;* D) *Beipiaosaurus* (modified from Xu, et al., 1999)*;* E) *Protarchaeopteryx* (after Ji, & Ji, 1997);   F) *Sinornithosaurus* (Dave) (modifed form Ji et al,. 2002); G) *Archaeopteryx* (modified from de Beer, 1954); H) *Microraptor gui* (modified from Xu et al., 2003)*;* I) *Cryptovolans* (modified from Czerkas, et al., 2002) ; J) modern bird. (not drawn to the same scale).

123

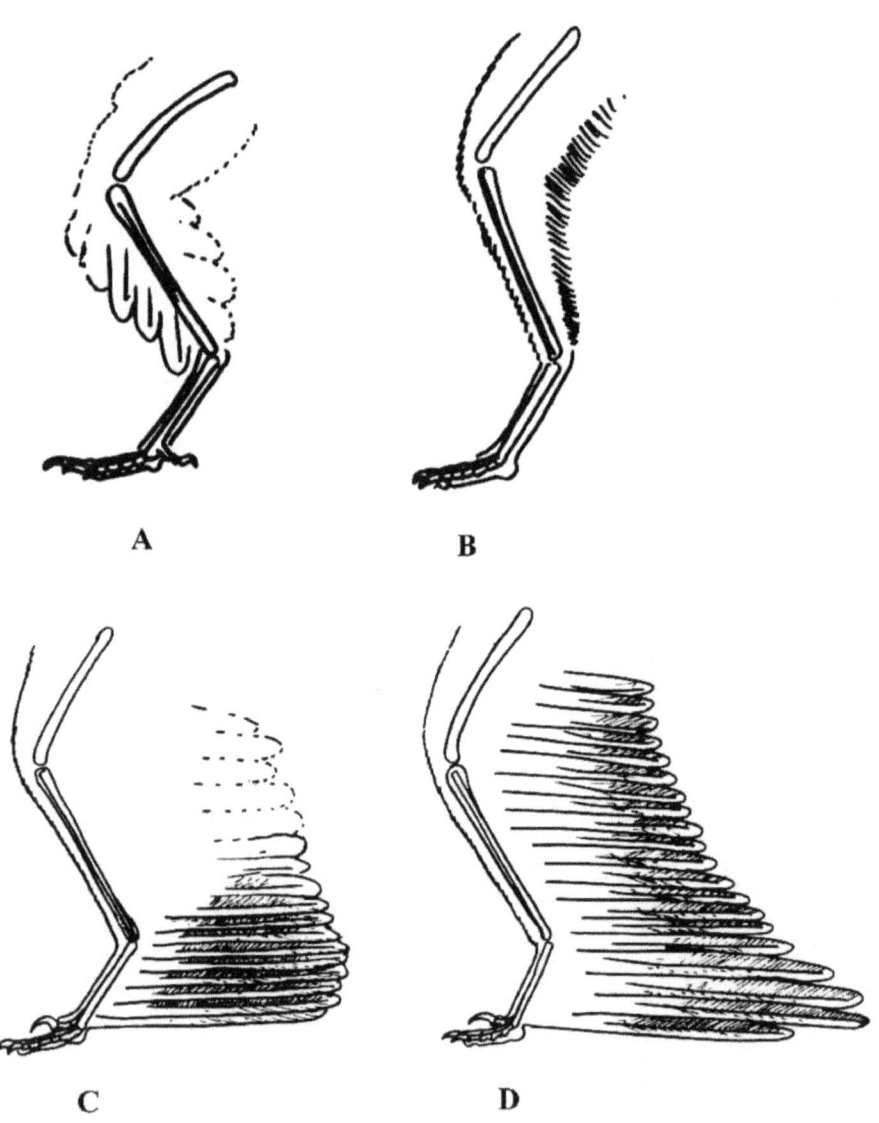

Figure 3). Leg feathers ; A) *Archaeopteryx;* B) *Sinornithosaurus* (Dave) (modifed form Ji et al,. 2002); C) *Microraptor gui* (modified from Xu et al., 2003); D) *Cryptovolans* (illustrated from the type and referred specimen). (not drawn to the same scale).

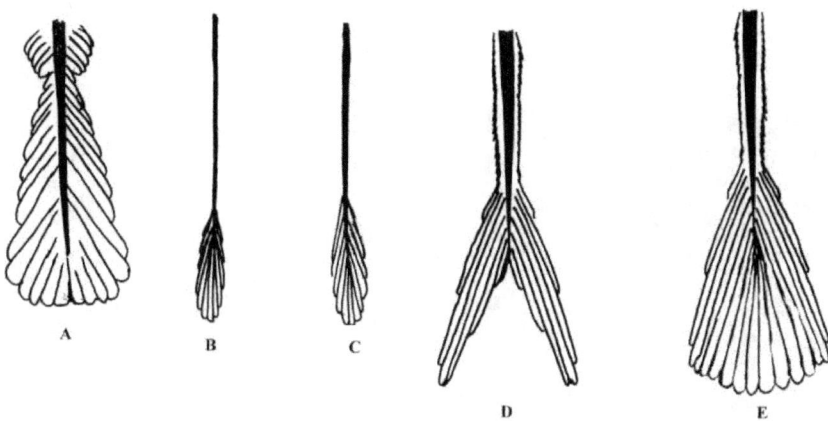

Figure
4). Tail feathers. A) *Archaeopteryx* (modified from de Beer, 1954); B) *Cryptovolans* (modified from Czerkas, et al., 2002) ; C) *Microraptor gui* (modified from Xu et al., 2003); D) *Caudipteryx* (modified from Ji, et al., 1998); E) *Caudipteryx* showing possible tail fan (modified from Ji, et al., 1998) (not drawn to the same scale).

Bibliography

de Beer, G. 1954. Archaeopteryx lithographica. A study base upon the British Museum specimen. Printed by order of The Trustees of the British Museum, Publication,. 224, 68pp.

Chiappe, L. M., Ji, S.-A. Ji, Q. and Norell, M. A. 1999. Anatomy and systematics of the Confuciusornithidae (Theropoda : Aves) from the Late Mesozoic of northeastern China: Bulletin of the American Museum of Natural History, 242, pp. 1-88.

Czerkas, S. A., and Yuan, C., 2002, An arboreal maniraptoran from northeast China: In: Feathered Dinosaurs and the origin of flight, edited by Czerkas, S. J., The Dinosaur Museum Journal, v. 1, p. 63-95.

Czerkas, S. A., Zhang, D., Li, J., and Li, Y., 2002, Flying Dromaeosaurs: In: Feathered Dinosaurs and the origin of flight, edited by Czerkas, S. J., The Dinosaur Museum Journal, v. 1, p. 97-126.

Ji, Q., Currie P. J., Norell M. A., and Ji S.-A., 1998, Two feathered dinosaurs from northeastern China: Nature, v. 393, p. 753-761.

Ji, Q., and Ji Ji S.-A., 1996, On discovery of the earliest bird fossil in China and the origin of birds: Chinese Geology, 1996, v. 10, n. 233, p. 30-33.

Ji, Q., and Ji S.-A., 1997, *Protarchaeopteryx*, a new genus of Archaeopterygidae in China: Chinese Geology, 1997, v. 3, n. 238, p.38-41.

Ji, Q., Ji, S.-A., Yuan, C.-X., and Ji, X.-X., 2002, Restudy on a small dromaeosaurid dinosaur with feathers over its entire body: Earth Science Frontiers, v. 9, n. 3, p. 57-63.Proctor, N. S., and Lynch, P. J., 1993, *Manual of ornithology, avian structure & function. Yale University Press, 340pp.*

Xu, X., Tang Z., and Wang X.-L., 1999, A therizinosauroid dinosaur with integumentary structures from China: Nature, v. 399, p. 350-354.

Xu, X., Zhou, Z., Wang, X., Kuang, X., Zhang, F., and Du, X., 2003, Four-winged dinosaurs from China: Nature, v. 421, p. 335-340.

Funny "Fossil" Bones · Humorous dino movies by Mark Berry

PREHISTORIC TIMES

#66 Jun/Jul 2004

Sunken Ships and lost fossils

The PT Interview:

Darren Tanke

The Dragon Ladies

FEMALE PALEOARTISTS

Great art, great articles, reviews and more....

Ford, T. L., 2004, How to Draw Dinosaurs. How theropods caught their prey, Part one, the skull: Prehistoric Times, n. 66, p. 18-19.

Chapter 20

How theropods caught their prey, part one, the skull

How theropods captured and ate their prey differs from clade to clade. This is due to several things; hands, claws, arms, neck, legs, teeth, length of skull, a solid or loose skull (kinesis) etc.

Teeth- Generally speaking the tooth morphology of theropods is very similar. The premaxilla and first dentary teeth are more robust while the side teeth are more blade-like. Tooth thickness, shape, and serrations differ enough to allow identification to at least a family level. Tyrannosaur premaxillary teeth are very distinctive. They are 'D' shaped with the serrations on the two edges of the 'D'. They have been found all the way back to the Late Jurassic. Allosaurus has a different style of 'D'shaped tooth with the serrations off set. Spinosaur teeth are round to oval. For the most part the very front teeth are smaller than the larger side teeth. In Spinosaurids the front teeth are larger than in other theropod clades. The front teeth are for nipping and the back blade like teeth were used for cutting.

Skull- Kinesis (kinetic) allows individual skull elements to move and not form a solid unmovable mass. Snakes are the extreme of this.The skull of *Allosaurus* is extremely kinetic and moveable. There have been many studies of this. The skull elements in Allosaurus are unfused (except for the braincase) (Figure 1). The snout can be raised or depressed, while the snout is lowered, the lower back part of the skull would slide back, and when the snout is raised the lower back part of the skull will move forward. The jaws can be expanded at the symphysis (where the jaws meet). This symphysis is unfused and there are muscles which allow the two sides of the jaw to separate (akin to a snake). The mandibular joint (where the dentary, surangular and angular meet which is the same all theropods) allows the jaw to slightly move up or down. The two sides of the skull can move slightly against each other, rising one side and lowering the other. In Bakker's book (The Dinosaur Heresies) he illustrates a *Ceratosaurus* with a kinetic skull (Figure 2). He talks about the lower half of the skull expanding which is done because of the loose attachment of bones. The quadrates would expand outward along with the quadrajugal, jugal, and maxilla. I'm not sure how kinetic the skull of *Ceratosaurus* is but I do know *Allosaurus* has a more kinetic skull. I am also not sure if the mandibular joint could flex outward as Bakker illustrated, though slightly up and down I can see. The nasals in *Allosaurus* are unfused but in ceratosaurs, abelisaurs, tyrannosaurs, spinosaurs the nasals are fused making their skull are more solid (akenetic). As I reported before, theropods for the most part have a narrow skull while *T. rex* has a wider skull. Peter Larson has expressed to me that *Tyrannosaurus* also had a kinetic skull though many are not convinced.

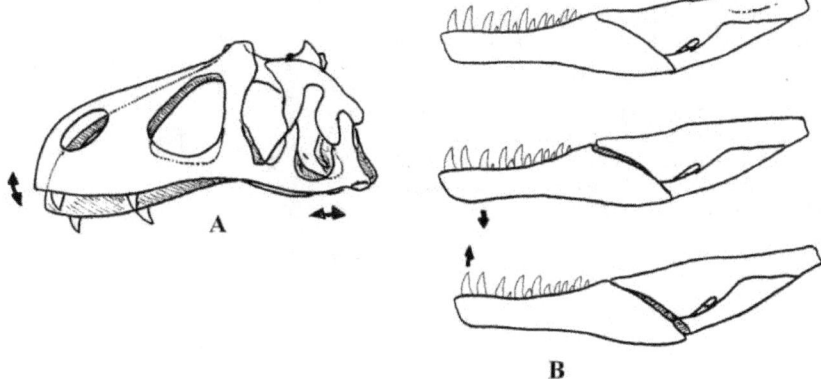

Figure 1:
Skull of *Allosaurus*, A) showing the kinetic skull moving up and down, and back and forth; B) lower jaw moving up and down.

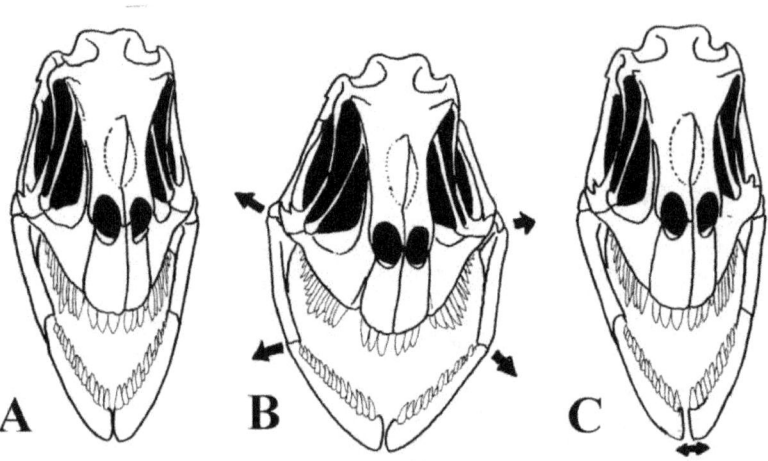

Figure
2): Skull of *Ceratosaurus* (after Bakker, 1986); A) solid skull; B) lower portion of jaw moving; C) just the symphysis moving.

Solid vs loose skulls-What is the difference? In snakes the skull is very mobile which allows them to expand their jaws to an incredible amount. The teeth point backwards and by moving the tooth bearing bones back and forth, a snake can 'push' its meal down its gullet. Theropods couldn't do this and didn't need to (Figure 3). The teeth in theropods do point backwards (for the most part) because it makes it harder for the prey to slip away. They can bite off chunks of meat big enough to swallow. Theropods, as noted before, can expand their lower jaws slightly to allow bigger chunks of food and the intermandibular joint may have allowed the jaw to manipulate the food enough to swallow (though this is speculation). The kinetic skull may have help to absorb the impact as the animal bit or captured its prey. There is no doubt that the prey was 'sliced' into smaller pieces to swallow by the blade like teeth. Most theropods were slicers, while *Tyrannosaurus rex* was a crusher! I disagree with Horner that *Tyrannosaurus rex* was a scavenger, but was an active hunter able to bite onto its prey and not let go and if small enough, crush it's prey in its mouth. Mike Triebold sells a cast of a *Triceratops* femur that has several theropod bite marks (*Tyrannosaurus rex* and dromaeosaurids) and holes from teeth (he calls it the bisket), and there is a *Triceratops* ilia that has a chunk bitten out. Recently in National Geographic a life size *Tyrannosaurus rex* skull was built (out of bronze I think) and simulated it crushing bones. It was very impressive. Also, in the SVP report I read a poster with a *Triceratops* horn being bitten off!! Now that is power!!!

The kinetic skull of *Allosaurus* didn't allow it to crush its prey. It was designed to have taken off chunks of meat in order to kill and could withstand side to side action or even a lot of skull movement of individual skull bones. Rauhut argues the long skulls of spinosaurids couldn't withstand the side to side torsion of catching large prey and just caught small prey with the tips of its jaws. I asked him if he believed they caught fish, and he was undecided on that. Tyrannosaurs could easily catch prey and take large chunks out, while *T. rex* could take bigger chunks and crush its prey.

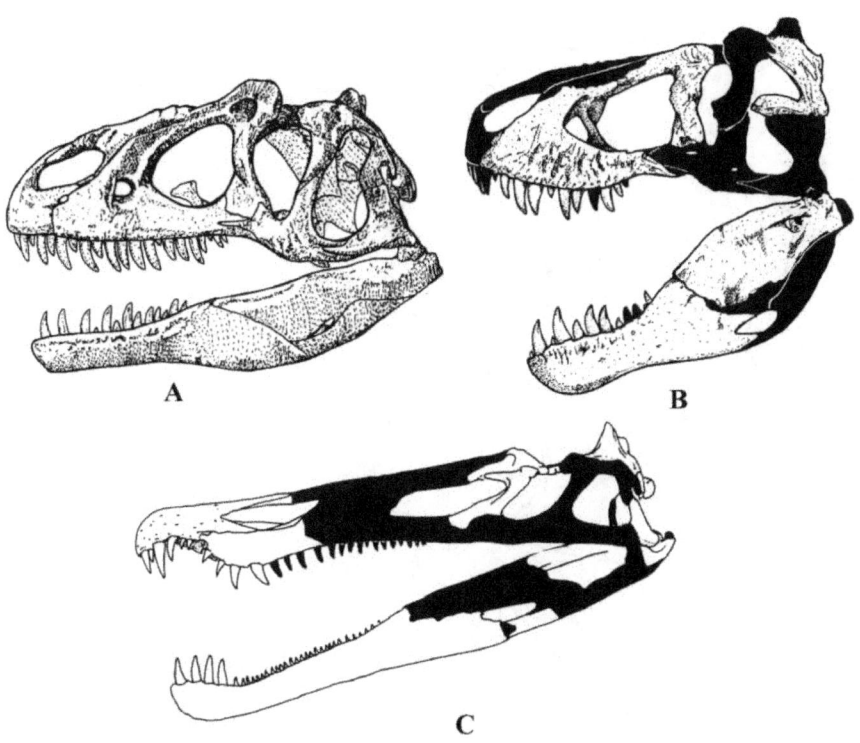

Figure 3): Skull of theropods. A) *Allosaurus* (a bitter) (after Madsen, 1976); B) *Tyrannosaurus rex* (holder and crusher) (modified from Osborn, 1905); C) *Baryonyx* (a nipper) (modified from Charig & Milner, 1997).

The forelimb- in theropods vary from the strong grasping arms of *Allosaurus* to the nearly useless arms of abeliasaurids (Figure 4). *Allosaurus* has large claws and hands used for grasping its prey while it took large chunks of flesh with its jaws. Megalosaurids also caught prey with their arms and jaws. Spinosaurids also have large arms and hands. *Baryonyx* has been argued to have caught fish with its large thumb claw, but I believe its long jaws would have been better suited for that. I'm not sure what the forelimbs of abelisaurids were used for, they are just too short and the hands couldn't hold prey.

Tyrannosaurs have small forelimbs, but they were still useful. Carpenter and Smith have studied the forelimb of *Tyrannosaurus rex* and found each arm was capable of holding several hundred pounds. Their research shows that the arms weren't atrophied and useless but were used to hold its struggling prey while the head dispatched it. They believe (as do I and I'm sure several if not all of you readers) that *T. rex* was an active predator and not a scavenger. They also believe *T. rex* was an ambush predator, which I disagree with.

In conclusion, *Allosaurus* would catch its prey with its hands and take chunks out with its jaws. Greg Paul and others believes they nipped large chunks of flesh from sauropods and waited for them to die, spinosaurs caught prey with the tip of its jaws, tyrannosaurs were active predators, killing with their massive jaws while the forelimb held on and *T. rex* was an active predator with crushing ability with its powerful jaws.

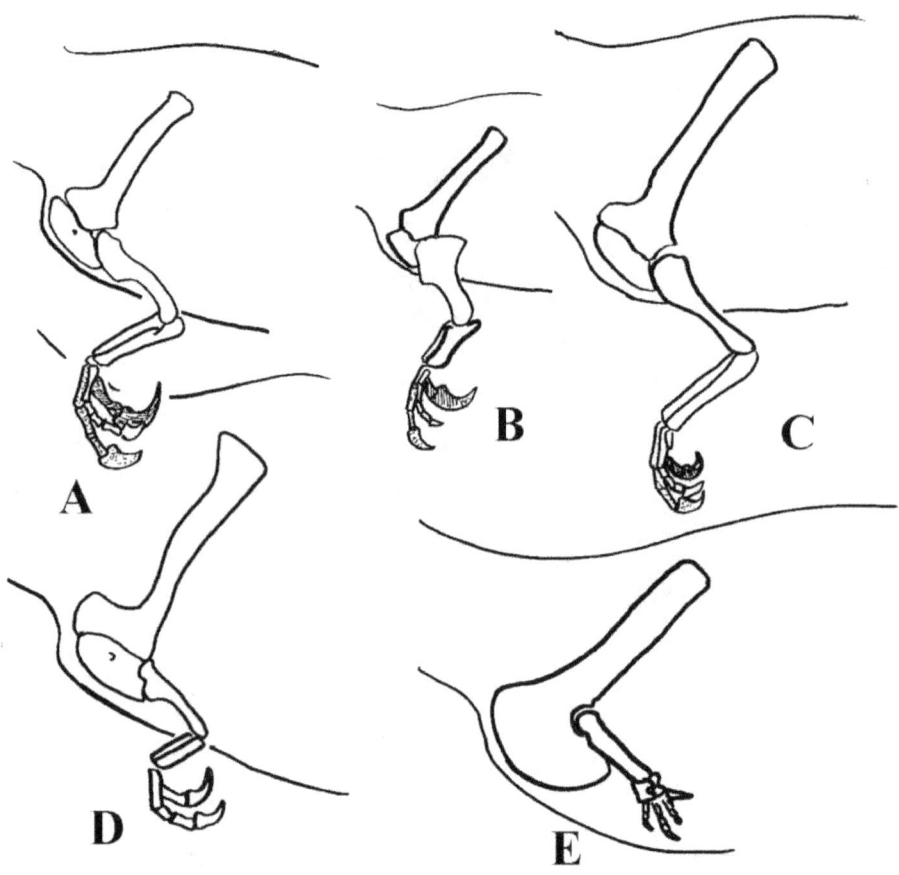

Figure 4); Arms of theropods; A) *Allosaurus* (after Gilmore, 1915); B) *Baryonyx* (after Charig & Milner, 1997); C) *Yangchuanosaurus* (after Dong, et al., 1978); D) *Tyrannosaurus rex* (after Carpenter & Smith, 2001); E) *Carnotaurus* (after Bonaparte et al., 1990).

Bibliography

Achenbach, J., 2003, Dinosaurs come alive: National Geographic, v. 203, n. 3, p. 2-33.

Bakker, Robert T. 1986. The Dinosaur Heresies, New Theories Unlocking the Mystery of the Dinosaurs and their extinction. William Morrow and Company, Inc. New York: 481pp.

Bonaparte, J. F., Novas, F. E., and Coria, R. A., 1990, *Carnotaurus sastrei* BONAPARTE, the Horned, Lightly Built Carnosaur from the Middle Cretaceous of Patagonia: Contributions in Science, n. 416, p. 1-41.

Carpenter, K., and Smith, M. B., 2001, Forelimb Osteology and biomechanics of *Tyrannosaurus rex*: In: Mesozoic Vertebrate Life, edited by Tanke, D. H., and Carpenter, K., Indian University Press, p. 90-116.

Charig, A. J., and Milner, A. C., 1997, *Baryonyx walkeri*, a fish-eating dinosaur from the Wealden of Surrey: Bulletin of The Natural History Museum, Geology Series, v. 53, n. 1, p. 11-70.

Dong, Z. M., Chang Y. H., Li X. M., and Zhou S. W., 1978, Note on a new carnosaur *Yanchuangosaurus* shangyuanensis gen. et sp. nov.) from the Jurassic of Yangchuan District, Szechuan Province: Kexue Tongabao, v. 5, p. 302-304.

Gilmore, C. W., 1915, On the fore limb of *Allosaurus fragilis*: Proceedings of the United States National Museum, v. 49, p. 501-503.

Madsen, J. H. jr., 1976, *Allosaurus fragilis* a revised osteology: Utah Geological and Mineral Survey, Bulletin, n. 109, p. 1-163.

Osborn, H. F., 1905, *Tyrannosaurus* and other Cretaceous carnivores Dinosaurs: Bulletin of the American Museum of Natural History, v. 21, p. 259-265.

PREHISTORIC TIMES

#67 Aug/Sep 2004

THE PT INTERVIEW:
JACK HORNER

New Utah Field House

The Art of
John Sibbick

JURASSIC PARK'S
JOHN BELL

NEWLY DISCOVERED
DINOSAURS & MUCH
MORE

U.S. $5.95 • Canada $6.95

Front cover art ©2003 Robert Walters & Paul Sorton

Ford, T. L., 2004, How to Draw Dinosaurs. A jab to the ribs: Prehistoric Times, n. 67, p. 18-19.

Chapter 21

A jab to the ribs

The idea for this issue's article stems from a post by Steven Coombs on the DML (Dinosaur Mailing List) on the internet. I haven't been on the list for quite a while now and I won't go into details why, suffice it to say I can still read the archives which is fine by me. Steven was asking about my PT article on barrel-like rib cage of *Spinosaurus* (issue 50, chapter four of this book). Ken Carpenter commented that there is no evidence for this and to take my reconstruction with a grain of salt. I had a few emails back and forth with Ken I thought the result of that would make a good PT article, so here it is.

I based my original assumption on the ribs which are oval and I thought the body cavity was barrel shaped and shallow. Ken pointed out that the ribs are from the posterior region of the body, thus shorter than the anterior ribs. I also thought if the ribs were angled back the ribs would dig into the body. Ken said I should look at Chris Brochu's monograph on 'Sue' because *T. rex* also has these barrel shaped ribs, yet the ribs over lap the proceeding vertebrae and don't dig into the body.

One of the things I like about writing these articles is that sometimes I'll learn something also. Now I know I've been guilty of not following how the ribs articulate and hopefully after writing this I'll remember to illustrate the ribs correctly.

The ribs in dinosaurs overlap the proceeding vertebra, but why do they do that? Is it the rib or the vertebra that determines this? Actually both. The curvature and shape of the rib and the transverse process of the vertebra.

A rib consists of the capitulum, tuberculum and shaft (Figure 1). The rib attaches to the transverse process (the wing like projection) via the tuberculum and the parapophosis (on the side of the centrum) via the capitulum (Figure 1). If the transverse process is perpendicular to the vertebra then the ribs do not overlap the proceeding vertebra. This is true for 'primitive' reptiles; anapsids, lizards, crocodiles, diapsids, etc. For dinosaurs and mammals the transverse process angles backwards and the rib overlaps the proceeding vertebra. This means the ribs should be illustrated overlapping the proceeding vertebrae in dinosaurs (Figure 2).

The ribs are not firmly attached to the vertebrae and can (and do) move slightly. The ribs curve out and back and act like a bellows in order for the animal to breathe.

Back to *Spinosaurus*. I illustrated the rib in perpendicular position to the vertebrae, which is wrong. After correcting the angle of the transverse process the rib angles back and out. By doing this, in dorsal view I can then determine how the vertebrae and ribs will look in side view. The rib cage becomes shallower and is a more typical theropod rib cage and does not look barrel shaped (Figure 3).

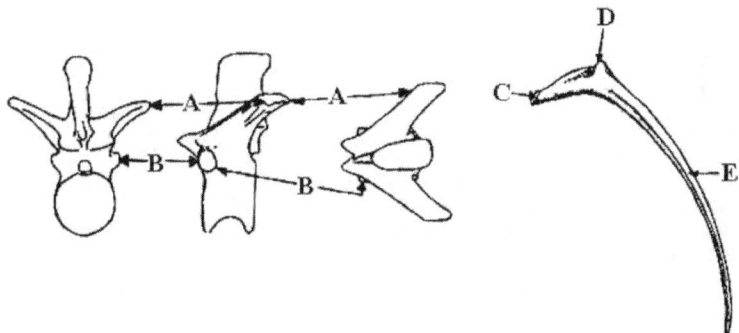

Figure 1): Vertebra and rib of *Allosaurus* (after Madsen's *Allosaurus* monograph, 1993); A) transverse process; B) parapophosis; C) capitulum; D) tuberculum and C) shaft.

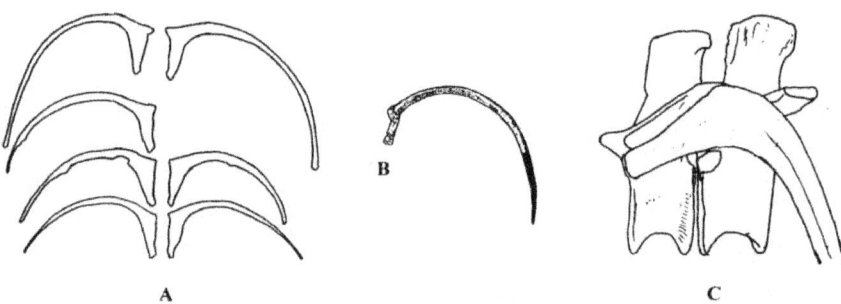

Figure 2): Ribs of A) *Tyrannosaurus rex* ('Sue' from Brochu, 2003) and B) *Spinosaurus* (after Stromer, 1915) showing how both have barrel-shaped ribs; and C) showing how the ribs overlap the proceeding rib in *Allosaurus*.

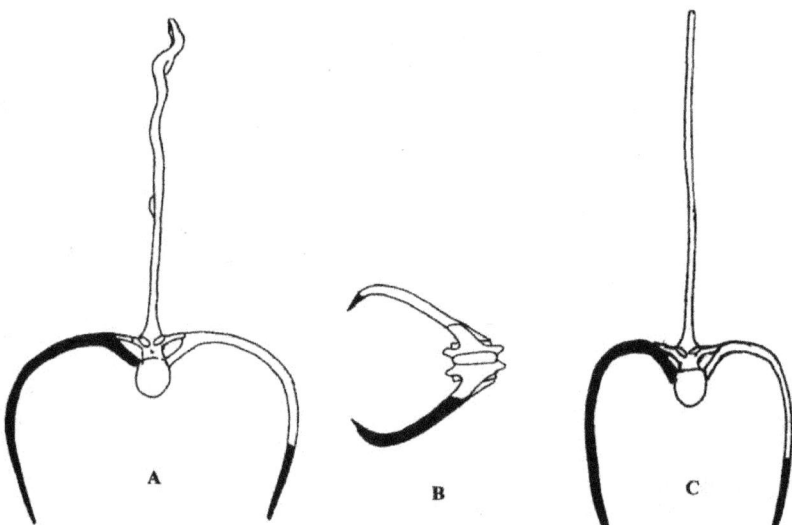

Figure 3): A) my original illustration of *Spinosaurus* showing the 'barrel-like' chest; B) dorsal view of ribs showing the corrected angle of ribs due to the angle of transverse process and C) corrected chest showing a more typical theropod like chest cavity. Also, note in figure A the tip of the spine is wavy, this is more than likely due to the plasticity that the bone goes through during fossilization and is not indicative the shape of the actual vertebra.

Bibliography

Brochu, C. A., (2002) 2003, Osteology of Tyrannosaurus rex: insights from a nearly complete skeleton and high-resolution computed tomographic analysis of the skull: Journal of Vertebrate Paleontology, v. 22, supplement to n. 4, memoir 7, 138pp.

Madsen, J. H. jr., 1976, Allosaurus fragilis a revised osteology: Utah Geological and Mineral Survey, Bulletin, n. 109, p. 1-163.

Stromer, E., 1915, Ergebnisse der Forschungsreisen Prof. E. Stromers in den Wusten Agyptens. II. Wirbeltier-Reste der Baharije-Stufe (unterstes Cenoman). 3. Das Original des Theropoden Spinosaurus aegyptiacus: Abhandlungen der Koniglich Bayerischen Akademie der Wissenschaften Mathematisch-physikalische Klasse 28, band 3, p. 3-32.

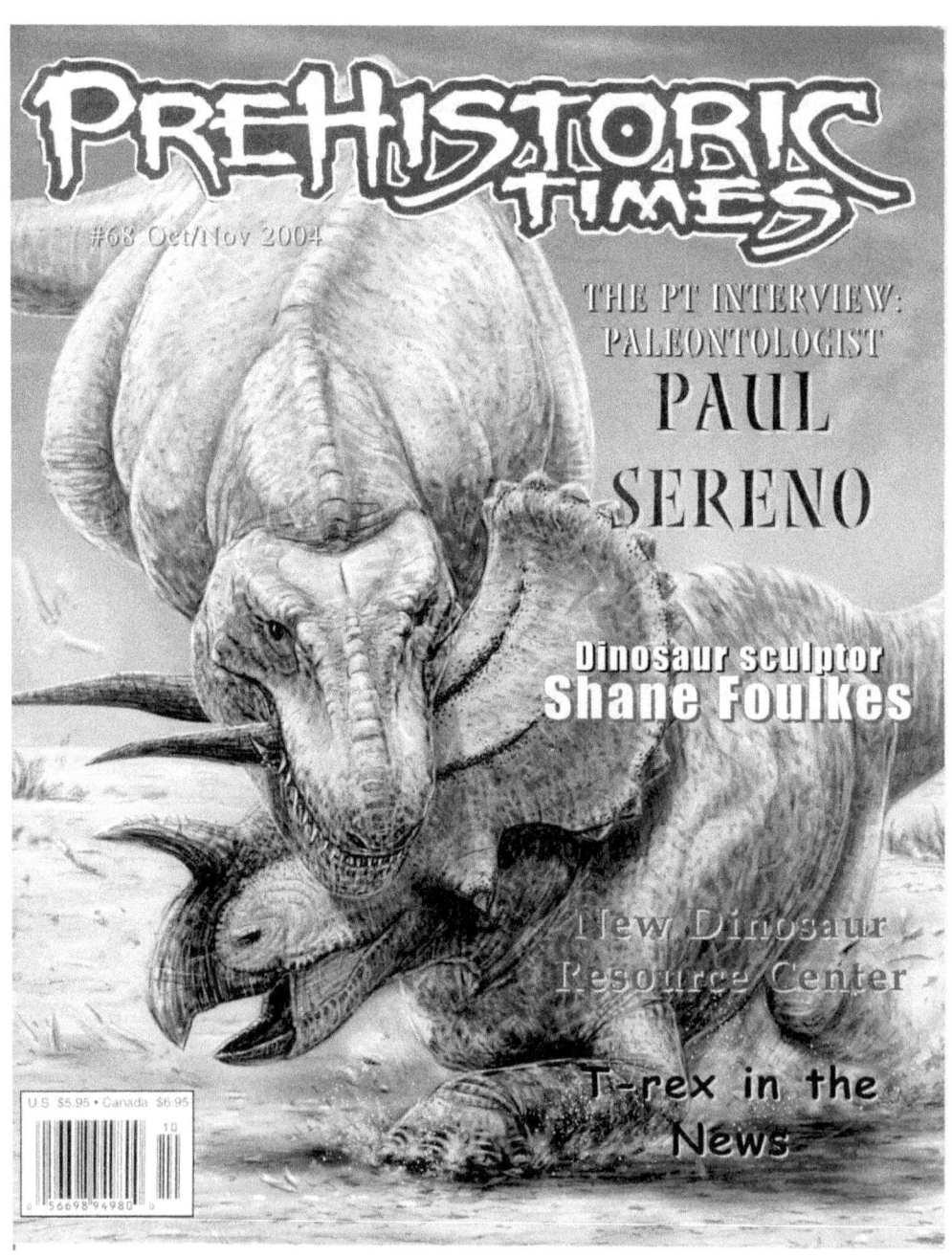

PREHISTORIC TIMES

#68 Oct/Nov 2004

THE PT INTERVIEW:
PALEONTOLOGIST
PAUL
SERENO

Dinosaur sculptor
Shane Foulkes

New Dinosaur
Resource Center

T-rex in the
News

U.S. $5.95 • Canada $6.95

136

Ford, T. L., 2004, How to Draw Dinosaurs. Armoring sauropods: Prehistoric Times, n. 68, p. 18-19.

Chapter 22

Armoring Sauropods

When I saw that this issue's spotlight dinosaur was *Saltasaurus* I decided to write about this interesting dinosaur. Sure I already did an article on sails, spikes and armor of sauropods but I didn't get into a lot of detail of the armor itself. So when inspiration strikes, go with it.

The very idea of an armored sauropod before the 1980's was either fantasy or Science Fiction. But in 1980 Bonaparte and Powell reported on *Saltasaurus* for the first time and showed that it was armored. The osteoderms were associated with skeletal elements and because they were different enough from ankylosaurs showed that this association was not in doubt. The few osteoderms that were found were disc shaped, along with some pebbly skin impressions. The problem is no articulated specimen (of any titanosaur) has been found with the osteoderms in place, so the exacted location of the osteoderms on the body is pure conjecture for now. The first sauropod to have osteoderms associated with it is '*Titanosaurus madagascarensis*'. Only one disc shape osteoderm was found. It was argued for decades that that osteoderm didn't belong to a sauropod. But due to several osteoderm found in the last 20 years it is know known that some sauropods did have armor (i.e. osteoderm, scutes).

Just a quick note on *Saltasaurus* itself (Figure 1). It was a small animal (actually a dwarf for a sauropod) with the largest femur being only about 80cm, and probably stood about 160 cm or less than 2 meters at the hip!!!

And was only about 5-7 meters long (about 15-21 feet, not the 10-12 meters that others have said). And all known *Saltasaurus* are adult or subadult. The neural spines are short and the back would have been 'flat'. The ilia flare out and had a very wide body. The first color painting of an armored *Saltasaurus* was by Mark Hallet for George Olshevsky's Science Digest article in 1981. Also, Mark Hallet was the first artist (that I know of) to show a feathered ornithopod (*Fuglatherium*) in his Australian dinosaur fauna painting (which I think is second best to his *Mamemchisaurus* painting).

Titanosaurs, so far, are the only sauropods known to have had osteoderms. I say so far because this can change any time with new discoveries.

Titanosaurs have been found in North and South America, Europe, Asia, and Africa. But osteoderms haven't been found in North America (though there is a slight possibility of a fragmentary one from Texas. Editors note: It now has been confirmed that *Alamosaurus* did have osteoderms).

In 2001 I gave a poster at the SVP on sauropod osteoderms. In that I tried to classify the types of osteoderms because not all the osteoderms are the same. I'll be going over the classification here.

Four types of titanosaur armor can be recognized; 1- a large oval osteoderm or disc; 2- a oval fatter osteoderm or mound; 3- an oval to tear-drop shaped ostederm with a spine or bulb; and 4- pavement (Figure 2).

Disc (discoidal): Emend Le Leouf *et al.*, (1994) and Ciski, (1999), oval to slightly circular in outline, a cingulum on the margins of the osteoderm which can have a ridge (slight elevated area in the middle or a strong ridge) or smaller round osteoderms attached to the rim (as in *Saltasaurus*). The ventral side can have a double ridge or be slightly concave. The dorsal surface can be rough and vascular, or nearly smooth. The double ridge was on the bottom of the osteoderm and is speculated by several paleontologist to have sat over the neural spine. The single ridged ones would have had the ridge on the outside of the skin. This type of osteoderm would have been embedded in the skin itself, fairly deeply at that.

Bulb: Emend Le Leouf *et al.*, (1994) and Ciski, (1999); the bulb (here taken as a type of the osteoderm) consists of an oval to tear dropped shaped osteoderm with a spine (spine of Le Leouff *et al.*, 1994) and a 'root' or 'foot' (cingulum of Le Leouff, *et al.*, 1994). The spine can be short or fairly tall and pointed and the 'root' can be small or very large. This osteoderm may have sat more loosely or not as deeply imbedded in the skin as that of a disc.

Mound: A large oval osteoderm with a slightly convex ventral area and a rounded dorsal area. The surface is rough and vascular and the ventral edge can be concave. This osteoderm may have sat more loosely in the skin than a disc.

Pavement: Small round ossicles that form a 'solid?' sheet or large scales.

Placement of osteoderms: Only two specimens have been associated osteoderms with skeletal elements, and both are caudal vertebrae. One is a disc osteoderm and the others are bulb osteoderms. This may indicate that the tail was covered by the osteoderms. Bonaparte & Powell, suggest that the sacro-pelvic region was covered in small oval osteoderm (pavement) sheets. Le Loeuff et al. suggest that the spinous osteoderms (bulbs) were displaced in longitudinal rows across the shoulder region (as found in some stegosaurids and ankylosaurids) and the flat oval osteoderms were over the back. Titanosaurids have wide flaring ilia, short dorsal neural spines and wide ribs giving the animal a large area for osteoderms to attach. In ankylosaurids the sacro-pelvic area is large and solid and could form a solid 'pelvic shield', because sauropods are of a similar design it is possible that titanosaurids also had a 'pelvic shield'. The tear-drop-shaped bulb osteoderms have a level base and would need an attachment area that would accommodate it. If bulb osteoderm were along the tail, the widest longest area for the osteoderm to fit would have been along the sides of the tail (similar to Stegosaurus). If the bulb osteoderm were along the body, the back would have been the best area for the placement of the osteoderms. Mounds and bulbs may have sat loosely, or have not have been as deeply imbedded in the skin while discoidal would have sat much more deeply.

Some osteoderms of Saltasaurus are disc shaped some with a single dorsal ridge and concave or flat ventral side while others have two ridges ventral ridges. These later osteoderms ridges were on the ventral side of the osteoderm and it has been speculated that the osteoderm sat over the dorsal neural spines with the ridges covering the dorsal tip of the neural spine. This is unusual because apart from titanosaurs all other armored dinosaurs; pseudosuchians, crocodiliformes etc, have a pair of osteoderms running down the midline of the body. This is why every illustration of Scutellosaurus showing only one row of spines down the back is wrong. Even Colbert's monograph states a right and left osteoderm, but I digress.

But how was the armor placed? (Figure 3). Just down the middle of the back? In two rows as Powell has suggested or all over the body as Mark Hallet showed? Not all the osteoderms are the same and were placed differently over the body. For now it is any ones best guess.

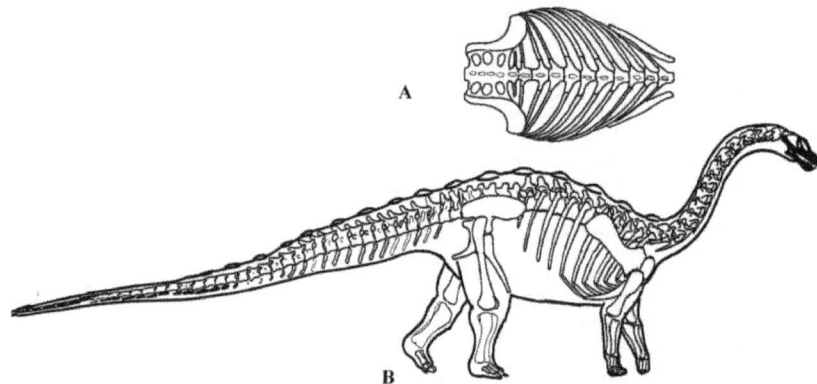

Figure 1). Skeleton of *Saltasaurus*; A) Dorsal view of skeleton; B) side view of skeleton.

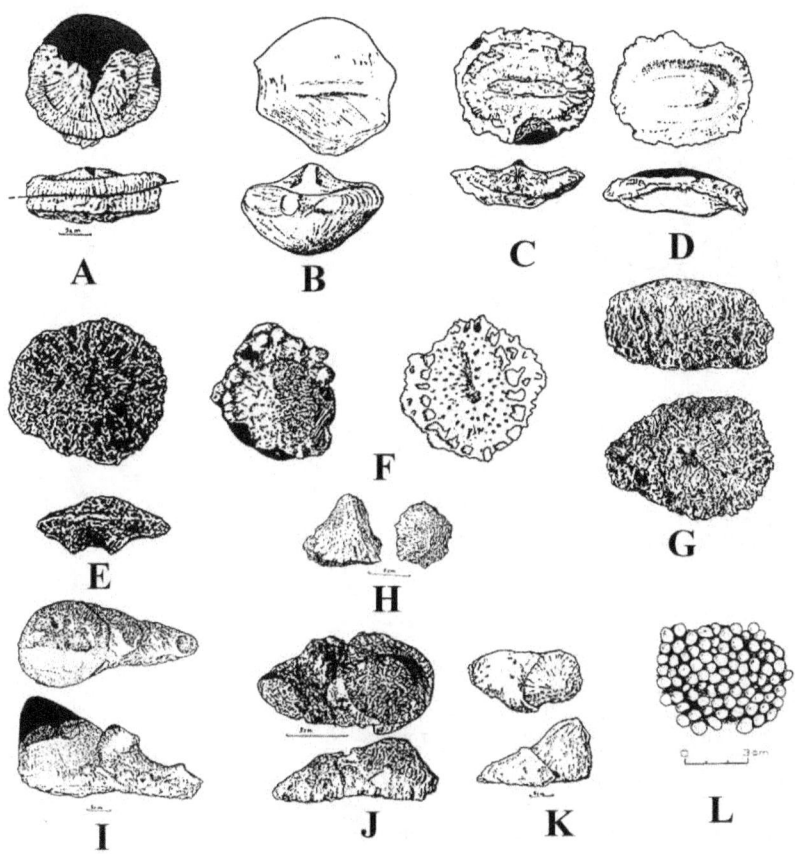

2) Titanosaur osteoderms; A-F Disc, G Mound, H-K Bulb and L pavement. A) *Titanosaurus madagascarensis* with a flat ventral side (after Deperet, 1986) ; B-D) *Loricosaurus scutatus* with a central ridge (after Huene, 1929), B having a concave ventral side, and C and D having a flat ventral side; E) osteoderm from an unnamed titanosaur from Brazil showing two ventral ridges (after Azevda & Kellner, 1998); F) *Saltasaurus* disc osteoderms with small pavement osteoderms along part of the rim (after Bonaparte, & Powell, 1980); G) Mound from an unnamed titanosaur from Europe (after Dodson, et al., 1998); H) and I) Bulb osteoderms from an unnamed titanosaur from South America (after Powell, 1986); J) Blub osteoderm from an unnamed titanosaur from Europe (after Le Leouff *et al.*, 1994); K) Bulb osteoderm from *Ampelosaurus* (after Le Leouff *et al.*, 1994); and L) Pavement osteoderms from *Saltasaurus* (after Bonaparte, & Powell, 1980).

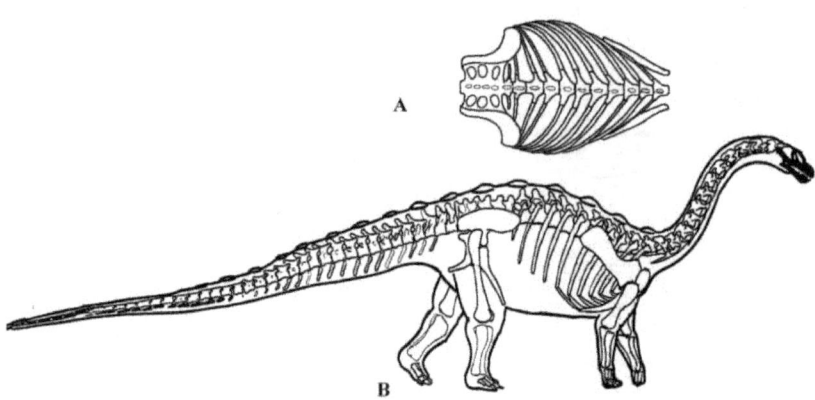

Figure
3). Illustrations of the position of osteoderms; A) Single row of osteoderms down the middle of the back and 'pelvic shield' made up of the 'pebbly' osteoderms (it is also possible the others had this also, or patches of 'pebbly' osteoderms); B) Double row of osteoderms down the back and along the side of the tail. Along the tail are 'bulb' osteoderms; C) Osteoderms in no discernable pattern (as per Mark Hallet, via Olshevsky, 1981).

Bibliography

Azevedo, S. A. de, and Kellner A. W. A., 1998, A Titanosaurid (Dinosauria, Sauropoda) osteoderm from the Upper Cretaceous of Minas Gerais, Brazil: Boletim do Museum Nacional, Geologia, v. 44, p. 1-6.

Bonaparte, J. F., and Powell J. E., 1980, A continental assemblage of tetarpods from the Upper Cretaceous beds of El Brete, northwestern Argentina (Sauropoda-Carnosauria-Aves). Ecosystemes Continentaux du Mesozoique: Memoires de la Societe Geologique de France, new series, n. 139, p. 19-28.

Ciski, Z., 1999, New evidence of armored titanosaurids in the Late Cretaceous – *Magyarosaurus dacus* from the Hateg Basin (Romania): Oryctos, v. 2, p. 93-99.

Deperet, C., 1896, Note sur les dinosauriens sauropodes et theropodes du Cretace superieur de Madagascar: Bulletin de la societie geologiques de France, v. 3, n. 24, p. 176-194.

Dodson, P., Krause, D. W., Forster, C. A., Sampson, S. D., and Ravoavy, F., 1998, Titanosaurid (Sauropoda) osteoderms from the Late Cretaceous of Madagascar: Journal of Vertebrate Paleontology, v. 18, n. 3, p 563-568.

Huene, F. von, 1929, Los Saurisquios y Ornithisquios de Cretaceo Argentino: Anales Museo de La Plata, series 2, n. 3, p. 1-196.

Le Loeuff, J., Buffetaut, E., Cavin, L., Martin, M., Martin, V., and Tong, H. 1994. An armoured titanosaurid sauropod from the Late Cretaceous of Southern France and the occurrence of osteoderms in Titanosauridae. p. 155-159. In Lockley, M. G., Santos, V. F. dos, Meyer, C. A., and Hunt, A. (ed), Aspects of Sauropod Paleobiology, Revista de Geociencias, Gaia, 10.

Olshevsky, G., 1981, Dinosaur Renaissance: Science Digest, v. 89, n. 7, p. 34-43.

Powell, J. E., 1986, Revision de los Titanosauridos de America del Sur: Universidad Nacional de Tucuman Facultad de Ciencias Naturales, Unpublished thesis, 340pp.

PREHISTORIC TIMES

#69 Dec/Jan 2005

Paul Sereno Part II

Stephen Czerkas

Nests, Eggs & Baby Dinos

Sculptor Tony McVey & Much More!!

U.S. $5.95 • Canada $6.95

Ford, T. L., 2004, How to Draw Dinosaurs. Beak heads? Theropods with keratin skulls: Prehistoric Times, n. 69, p. 18.

Chapter 23

Beak heads? Theropods with keratin skulls.

For the past decade or so, several paleontologist (Currie, Horner, Witmer, Larson, etc) have toyed with the idea that some theropods had a "horny or keratin' covering on their skull. I'm not talking about the mouth area (though there are two ornithimimids that do show keratin around the tip of the jaws) but on top of the skull; i.e. the nasals, frontals, lachrymals, and postorbital. They argue that it is the theropod with a rugose or striated bones that allowed the keratin to adhere to.

Allosaurus may have had keratin on the sides of the nasals (which are paired and loosely attached to each other) and the lachyrmals. *Ceratosaurus* may have had keratin on its 'horns' which would increase the size of the 'horns'. *Dilophosaurus* may have had keratin over its head frill, which would have strengthened the frill. Sinraptorids, *Monolophosaurus*, Carcharodontosaurids, Abelisauroidea, and tyrannosaurids may all have had keratin. *Carnotaurus* could have had keratin horns as well as the nasals and *Majungatholus* had a keratins single 'bump' and nasals. In *Tyrannosaurus rex* there is both slightly rugose to heavily rugose nasals. I am fascinated by this and I can't tell you how many times I've photographed the cast of AMNH 5027 in as many angles that I could. Incidentally there are small sand grain sized 'bumps' on the left (?) side postorbital which may be skin, though Mark Norell has assured me that it isn't. Over the orbit in *Tyrannosaurus rex* there has now been found an additional bone. In 'Sue' this is fused to the skull, but in 'Stan' it was found years later as float. I remember fondly when Neal Larson showed me that small funny looking bone. This may also have had a keratin covering. Some artist like putting large scales around the eyes. There is no proof of this and for now it is up to the artist whether or not to put them there.

What would the keratin have looked like? Smooth and shiny or rough and dull? Was it a hard keratins covering like claws and beaks or soft like that of a crocodilian scute? I believe the older animals would have had a more worn and dull look while the 'youngins' would have been smoother and 'cleaner' looking.

What would the keratin been used for? To strengthen the skull? Sexual dimorphism? For protection in combat? All are possible.

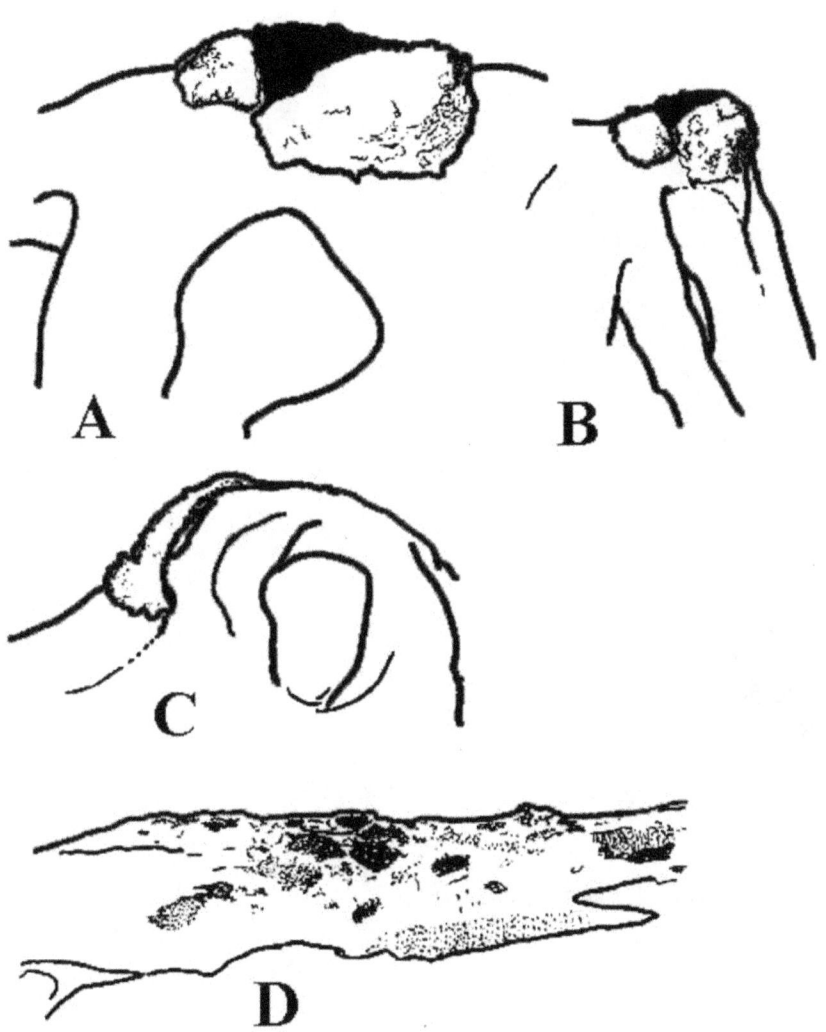

Figure 1). *Tyrannosaurus rex* (Stan) skull elements. A-C) New bone over the orbit. A) Side view; B) Front view; C) Top view; D) Nasal showing the rugose structure.

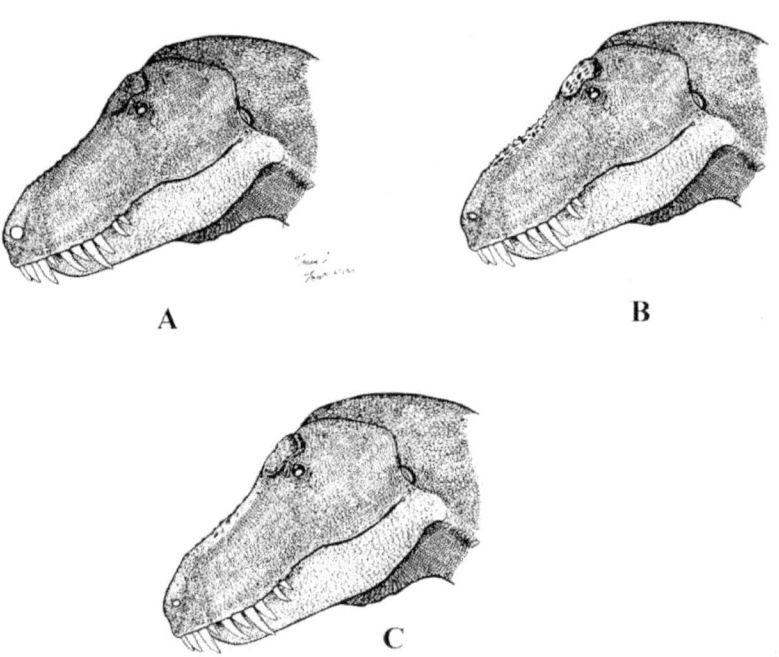

Figure 2). 'Stan' 3 different possibilities; A) Without keratin; B) Smooth keratin; C) and Rough keratin.

PREHISTORIC TIMES

FEB _____ #70

PT INTERVIEWS:
SCULPTORS
STEPHEN CZERKAS
PART TWO – PLUS
BRIAN COOLEY

T-REX IN THE MOVIES
CHRIS BENNETT ON
PTEROSAURS

—AND MUCH MORE!

©2005 Wm Stout

US $5.95 • Canada $6.95

02

0 56698 94980 0

145

Ford, T. L., 2005, How to Draw Dinosaurs. Cheeky Ornithopods? (Part 1) Prehistoric Times, n.70, p. 18-19.

Chapter 24

Cheeky Ornithopods? (Part 1)

This is my 50th article of How to Draw Dinosaurs. I never really thought about getting here when I started but I kept writing doing my best not to miss an issue. Mike keeps this great mag going so I keep writing. I thought the topic for the 50th article should be a bit 'meat' to it than usual and because of that, it will be a two part article.

When I grew up, dinosaurs were dim-witted, slow moving, swamp dwelling animals. Sure they looked cool and my imagination would run wild but when the dinosaur renaissance of the 1970's came about I was a bit skeptical about the new views. Those old views were hard to change. In 1972 Peter Galton wrote an article about ornithischians with cheeks. At the time I had a hard time believing it and it took years for me to come around. Then about 5 years ago this view was again challenged and I thought about it and changed my mind.. Stephen Czerkas (1998) and Larry Witmer (1998), and Michael Papp and Larry Witmer (1998) have independently reached a conclusion that challenges this view. I was writing an article about ornithischians not having cheeks and sent rough drafts to those working on the subject. The paper wasn't finished due to conflict of research and I hadn't done any more work on si nce then but I'm glad I didn't publish the paper because I've again changed my mind. For more than two decades this view has dominated, but now that view is being challenged.

Galton wasn't the first to theorize ornithischians with cheeks. In 1903 Lull, working on Triceratops, wrote "Food gathered by the cutting beak was probably chopped into short pieces by the teeth, being kept in the mouth by the muscular walls of the cheeks. It is doubtful whether the gape of the mouth had a posterior extent further than the anterior end of the tooth series, as otherwise the portions of food chopped off, falling outside of the lower jaws, could not be retained in the mouth". Later research by Lull extended the cheeks to hadrosaurs. Others restored certaopians with cheeks (Lull & Wright, 1942, L. S. Russell, 1935, Sternberg 1951 and Eaton, 1960). But this was challenged by Brown & Schlaikjer in 1940. They argued that reptiles don't have the buccinator muscles like mammals or facial muscles. Haas (1955) no sauopsid reptile had cheeks and he reconstructed the jaw musculature with then. Cheekless ornithischians had been wildly accepted until the early 70's.

The Skull in ornithischians. First we'll look at the skull of *Hypsilophodon* (Figure 1) to understand the skull elements. The early ornithischians (fabrosaurs) have slightly inset buccal region, while most ornithischians (hadrosaurs, ceratopians and ankylosaurs) have deeply inset teeth to various degrees of development. It is because of this inset buccal region that Galton and others have argued for cheeks (Figure 2).

Both the premaxilla and predentary had a keratinous covering (see PT 53, chapter 7 of this book). The predentary fits inside the premaxilla and formed a 'cutting or shearing area'. Several ornithischian specimens have crenellations or bumps on the premaxilla and predentary and the horny sheath would follow this basic pattern. These crenellations would help in food gathering by scrapping off the vegetation. This area was unable to have served for an attachment area for lips and they wouldn't have been able to snarl. Some ornithischians had teeth in the premaxillary region (fabrosaurids, heterodontosaurids, hyspilophodontids and pachycephalosaurians).

As I pointed out in PT 53, the ornithischian beak varies in width. Fabrosaurids, heterodontosaurids, hypsilophodontids has a 'pointed' narrow beak. Camptosaurids, thescelosaurids and iguanodontids have wider shallow beaks (see PT 53). Ceratopian's have a thin pointed-beak. Hadrosaurs have the largest size variance of beaks with *Telmatosaurus* and other primitive hadrosaurs having a shallow beak or bill, with *Edmontosaurus* having the widest bill.

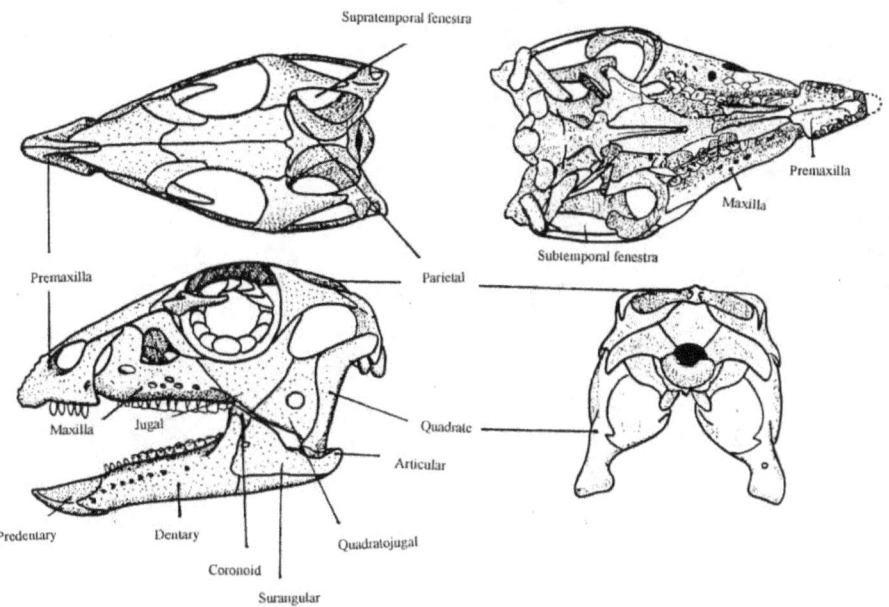

Figure
1). Skull of *Hypsilophodon foxi* in side, ventral, dorsal (after Galton, 1974) and caudal view (after Weishampel, 1984) showing the premaxilla, maxilla, jugal, quadrate, quadratojugal, predentary, dentary, coronoid, surangular, articular, subtemporal fenestra, supratemporal fenestra, and parietal.

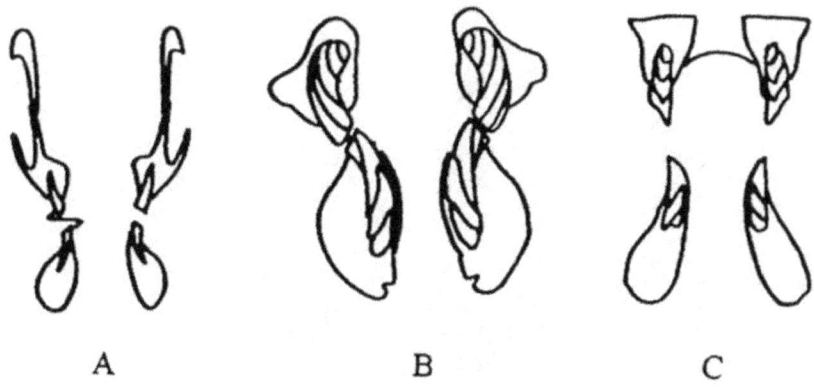

Figure
2). Transvers section of A) *Hypsilophdon*; B) *Edmontosaurus*; and C) *Triceratops* showing the buccal region (after Galton, 1972).

147

Closing the jaws. An important aspect in understanding if ornithischians had cheeks is to understand how the mouth closed. The back teeth are the first to occlude and continue to 'cut', not mash, the food until the lower jaw is completely closed. The jaw opens and closes at the cranial-mandibular joint (Figure 3). The lower jaw differs from the basal ornithopods and the more derived ones. In basal ornithischians (fabrosaurids, hypsilophodontids, thescelosaurids, camptosaurids, psittacosaurs and protoceratopians) the coronoid. In *Hypsilophodon* the teeth end in the middle of the coronoid process while in iguandontids, hadrosaurids and ceratopians the coronoid has transformed into an 'arm' that separates from the posterior mid section of the dentary which allows the tooth row to extend beyond the coronoid. The tooth row of *Iguanodon* (Norman, 1980) begins to extend into the coronoid area, while with hadrosaurids and ceratopians, the tooth row extends to the edge of the back of the coronoid (Figure 4 B, F, G. H).

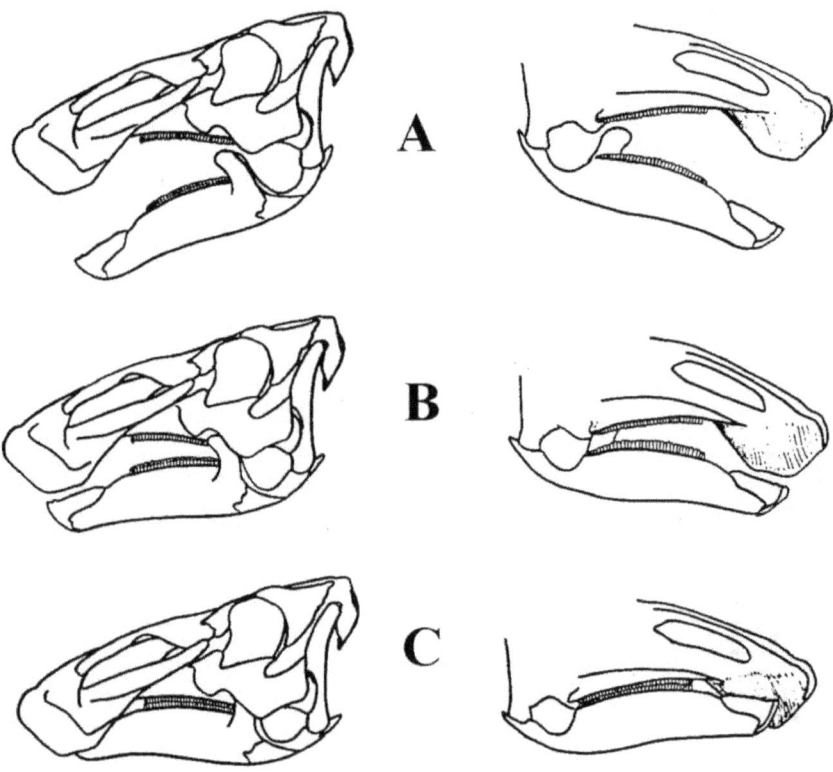

Figure 3). Closing the jaws of *Edmontosaurus* in lateral and internal views. A) open mouth: B) partially closed mouth, and C) closed mouth.

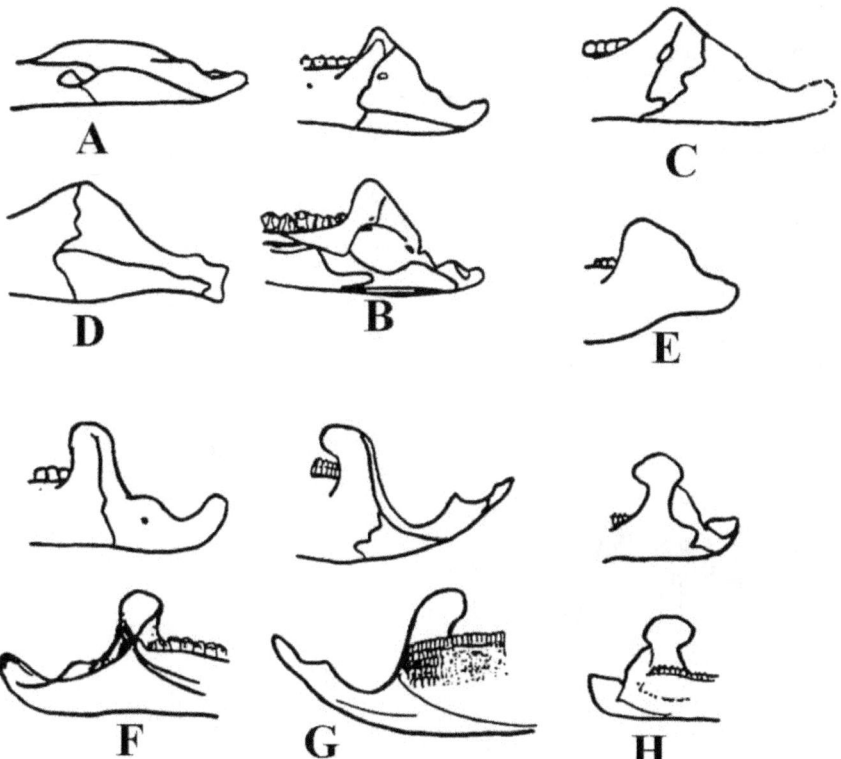

Figure

4). Side view of coronoid process of ornithischians. A) Fabrosaurid *Lesothosaurus diagnosticus* (after Sereno, 1991); B) Hypsilophodontid *Hypsilophodon foxi* with internal view (after Galton, 1974); C) Thescelosaurid *Bugenasaura infernalis* (after Galton, 1995); D) Psittacosaurid *Psittacosaurus neimongoliensis* (after Russell & Zhou, 1996); E) Protoceratopsid *Protoceratops andrewsi* (after Brown, & Schlaikjer, 1940); F) Iguanodontid *Iguanodon atherfieldensis* with the internal views (after Norman, 1996); Hadrosaurid *Edmontosaurus regailis* with internal view (after Lull & Wright, 1942); H) Ceratopian *Triceratops* (*prosus*) *horrridus* with the internal view (after Hatcher, Marsh, & Lull, 1907).

As the jaw is closed the coronoid fits behind the jugal and into the subtemporal fenestra. The M. pseudotempralis muscle attaches to the top of the coronoid, passes through the suptemporal fenestra and out the supertemporal fenestra and attaches to the middle of the parietal. Not all the jaw muscles will be dealt with in this paper. The M. depressor mandibulae helps to close the jaw at the caudal end of the jaws. As the jaw continues to close, the upper and lower teeth occlude with the upper teeth. The finished cycle of the closing of the mouth has the predentary fitting into the premaxilla.

The teeth occlude differently in the clades of ornithischians. *Fabrosaurus* teeth that are in the jaws don't show wear facets but shed teeth had extreme wear. They also show interesting wear facets in that they occlude on the front of one tooth and the back of the other. Heterodontosaurids has an oblique tooth to tooth contact that has been interpreted as a grinding surface. Hypsilophodontids camptosaurs, thescelosaurs, dryosaurs, and iguanodontids have wear patterns that are perpendicular to each other and an increase in tooth number with tooth to tooth wear causing a shearing action. The teeth are tightly packed and form 'dental' batteries in hadrosaurs and ceratopians. While they both have tooth batteries, there are differences in how the teeth grow and occlude to one another between the groups (Figure 5). When the skull of hadrosaurs is sectioned transversely it shows that the teeth grow in slight angles toward

149

each other. The upper teeth angle inward and the lower teeth angle outward. The teeth do not form a horizontal chopping edge, but are at an oblique angle. There is also a very thin piece of bone that lies on the inside of the dentary covering the tooth batteries, so that the inside of the teeth were not used in eating. Also, the teeth in the teeth battery in hadrosaurs grows not in a single horizontal rob but at an oblique angle; the back teeth grow first then the next and so on. In ceratopians the teeth grow upward in a single row as well as a horizontal row. You'd think the most efficient way for the teeth to be arranged would be for the flat surface of the teeth to be angled toward the inside of the lower jaw, but it's the opposite, the smooth flat area is toward the outside of the skull. This may be a good indication for cheeks.

In Ceratopians the teeth have two roots (except for the basal ceratopian *Zuniceratops*). The tooth pushing out the upper tooth appears to have 'split' the upper tooth, thus making a double root. The teeth grow relatively straight with a perpendicular cutting area. A split root may be another way to give a stronger bite force while cutting its food.

The buccal areas of the lower and upper jaws vary in ornithischians. Fabrosaurids, heterodontosaurids, hypsilophodontids, thescelodontids and camptosaurids have shallow buccal areas, with iguanodontids, hadrosaurians and ceratopians having larger areas (Figure 2). It is because of this recessed area that cheeks are believed to occur.

Hadrosaurians have a large buccal area. A possible reason why the teeth grow in angles may have helped strengthen the cutting edge or grinding (Galton, 1972). This area is a small shallow area (in width but long in length) and would supply a very small area for grinding but long for shearing. To do this, the lower jaw would have to move forward and aft and up and down along the tooth area.

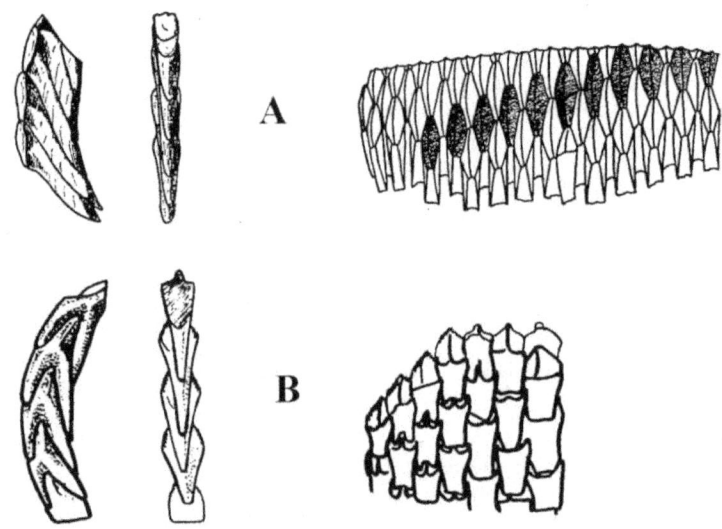

Figure 5). Teeth growth of the tooth batteries showing the difference in tooth growth and tooth batteries. Both genera would have a small thin piece of bone (that is rarely fossilized and the 'inside of the teeth would not have been used in eating) of A) *Edmontosaurus*, showing a single row of teeth in side, transverse section and tooth battery with darker teeth showing in angle in which the teeth grow (after Ostrom, 1961)); B) *Triceratops* showing the same but the teeth grow upward and not obliquely like in hadrosaurs (after Hatcher, Marsh, & Lull, 1907).

The buccal shelf area of ornithischians jaws have few small foramina (though rarely illustrated), and these foramina need to be accounted for. I argue that theropods have foramina which had nerves and veins coming out of them to supply the mouth region with a tactile touch area similar to alligators but would also lack lips to cover the teeth. Theropods have larger amount of foramina that extends up and down the premaxilla, maxilla and dentary, than ornithischians do. In ornithischians, I believe the foramina feed labial oral glands via veins and nerves (more on this in the next article).

Bibliography

Brown, B. B., and Schlaikjer, E. M., 1940, A new element in the Ceratopsian jaw with additional notes on the Mandible: American Museum Novitates, n. 1092, p. 1-13.

Czerkas, S., 1998, The lips, beaks, and cheeks of Ornithischians: Journal of Vertebrate Paleontology, v. 18, supplement to n. 3, Abstracts of papers. Fifty-eighth annual meeting, Society of Vertebrate Paleontology, Snowbird Ski and Summer Resort, Snowbird, Utah, September 30-October 3, p. 37a.

Eaton, T. H., 1960, A new armored dinosaur from the Cretaceous of Kansas: University of Kansas Palaeontological Contributions, Vertebrata, Article 8, 24pp.

Galton, P. M., 1972, Classification and evolution of ornithopod dinosaurs: Nature, v. 239, p. 464-466.

Galton, P. M., 1972, The cheeks of ornithischian dinosaurs: Lethaia, v. 6, p. 67-89.

Galton, P. M., 1974, The Ornithischian Dinosaur *Hypsilophodon* from the Wealden of the Isle of Wight: Bulletin of the British Museum (Natural History), Geological Series, v. 25, n. 1, p. 3-152.

Galton, P. M., 1995, The species of the basal hypsilophodontid dinosaur *Thescelosaurus* Gilmore (Ornithischia: Ornithopoda) from the Late Cretaceous of North America: Neües Jahrbuch für Geologie und Palaontologie, Abhandlungen, v. 198, n. 3, p. 297-311.

Haas, G., 1955, The Jaw Musculature in Protoceratops and in Other Ceratopsians: American Museum Novitates, n. 1729, p. 1-24.

Hatcher, J. B., Marsh, O. C., and Lull, R. S., 1907, The Ceratopsia: Monographs of the United States Geological Survey, v. 49, p. 1-300.

Lull, R. S., 1903, Skull of *Triceratops serratus*: Bulletin of the American Museum of Natural History, v. 19, p. 685-695.

Lull, R. S., 1908, The cranial musculature and the origin of the frill in the ceratopsian dinosaurs American Journal of Science, 4th series, v. 25, p. 387-399.

Lull, R. S., and Wright, N. E., 1942, Hadrosaurian Dinosaurs of North America: The Geological Society of America, Special Paper, n. 40, p. 1-242.

Norman, D. B., 1986, On the anatomy of *Iguanodon atherfieldensis* (Ornithischia: Ornithopoda): Bulletin del l'Instut Royal Des Sciences Naturelles de Belgique, Sciences de la Terre, v. 56, p. 281-372.

Ostrom, J. H., 1961, Cranial morphology of the Hadrosaurian dinosaurs of North America: Bulletin of the American Museum of Natural History, v. 122, Article 2, p. 37-186.

Papp, M. J., and Witmer, L., 1998, Cheeks, beaks, or freaks: a critical appraisal of buccal soft-tissue anatomy in ornithischian dinosaurs: Journal of Vertebrate Paleontology, v. 18, supplement to n. 3, Abstracts of papers. Fifty-eighth annual meeting, Society of Vertebrate Paleontology, Snowbird Ski and Summer Resort, Snowbird, Utah, September 30-October 3, p. 69a.

Russell, D. A., and Zhao, X.-J., 1996, New psittacosaur occurrences in Inner Mongolia: Canadian Journal of Earth Science, v. 33, p. 637-648.

Russell, L. S., 1935, Musculature and function in the Ceratopsia: Bulletin of the National Museum of Canada, v. 77, p. 39-44.

Sereno, P. C., 1991, *Lesothosaurus*, "Fabrosaurids," and the early evolution of Ornithischia: Journal of Vertebrate Paleontology, v. 11, n. 2, p. 168-197.

Sternberg, C. M., 1935, Hooded Hadrosaurs of the Belly River Series of the Upper Cretaceous Canadian Department of Mines, National Museum of Canada, Bulletion, n. 77, p. 1-23.

Sternberg, C. M., 1951, Complete skeleton of *Leptoceratops gracilis* Brown from the Upper Edmonton member on Red Deer River, Alberta: Bulletin of the National Museum of Canada, v. 123, p. 225-255.

Weishampel, D. B., 1984, Evolution of jaw mechanisms in ornithopod dinosaurs: Advances in Anatomy, Embryology and Cell Biology, v. 87, Springer-Verlag, Berlin, Heidelberg, New York, Tokyo, 116pp.

Witmer, L. M., 1998, Application of the extant phylogenetic bracket (EPB) approach to the problem of anatomical novelty in the fossil record: Journal of Vertebrate Paleontology, v. 18, supplement to n. 3, Abstracts of papers. Fifty-eighth annual meeting, Society of Vertebrate Paleontology, Snowbird Ski and Summer Resort, Snowbird, Utah, September 30-

PREHISTORIC TIMES

#72 June/July 2005

12 YEAR Anniversary issue

Interviews with:
Author Steve Alten
Paleontologist Ken Carpenter
Artist Raul Martin
PLUS
The JH Miller Story
&
Land of the Lost
and much more

U.S. $5.95 • Canada $6.95

152

Ford, T. L., 2005, How to Draw Dinosaurs. Cheeky Ornithopods? (Part 2) Prehistoric Times, n.72, p. 20-21.

Chapter 25

Cheeky Ornithopods? (Part 2)

Editors note: I write this with a little egg on my face. Last chapter I wrote that it was my 50th article, and that I had not missed an issue of Prehistoric Times. However, as I put this book together I noticed that what I thought was my 50th issue, was actually my 49th, and that when I numbered my articles I missed a number. Also, for personal reasons I missed the following issue (71), and my 50th article was in issue 72, not 70.

What do Mammal cheeks look like? And do they look like what dinosaurs did? The cheeks in mammals serve several functions; eating, storage and communication. The cheeks help keep food in the mouth during chewing and redistribution of food for mastication. Mammals also have fleshy prehensile lips that aid in food gathering and food manipulation. The cheeks in some mammals are also used as a storage pouch aiding in the gathering of large quantities of food, mainly in rodents. Mammals have a shallow buccal region, inset teeth, that corresponds to where the buccinator muscles attach. The cheek muscle (*M. buccinatoris*) in mammals originates along the side of the maxilla and dentary. The cheek muscles do not leave muscles scars on the maxillae or dentary in mammals (Galton, 1972).

Mammalian lips and cheeks also serve in communication. Communication is achieved by sound manipulation as air passes through the vocal chords, lips and cheeks. Another form of communication is by snarling, this is achieved when the lips and cheeks are pulled back to show teeth in an aggressive behavior.

Pro cheeks. Galton, 1972 argued the buccal area in ornithischians was for muscle attachment for cheeks. The buccal muscles have vertical fibers which attach to the upper and lower buccal area and like mammals don't leave muscle scars. Lull (1908) argued for cheeks in *Triceratops*, along with Russell, 1935 for *Chasmosaurus*, and Lull & Wright, 1942 for hadrosaurs. Galton's defense for cheeks is...."I believe that most ornithischians had cheeks and a small subterminal mouth as restored for *Hypsilophodon* (Galton, 1971a, Fig. 5). Ornithischian dinosaurs represent the largest adaptive radiation of non-mammalian terrestrial herbivores and I suggest that the development of cheeks was an important reason for the success of these dinosaurs..."

Dinosaurs didn't masticate their food like cows because their jaws couldn't move side to side in a grinding motion. Ornithischians cut their food with their teeth. The teeth are so tightly set that them form a shearing plain (like scissors) when they close their jaws. They moved their jaws up and down and forward and back in a cutting/slicing action.

The intake of food is done by the beak, which crops the food. It then moves through the 'shallowest' part of the mouth and into the mouth itself. The tongue is used to manipulate the food toward the teeth and then chopped/cut into smaller pieces then swallowed. The teeth occlude in hadrosaurs and ceratopians are offset toward the outside of the jaw so any scrap of food would actually fall outside of the mouth.

Ornithischians ate more food (vegetation) than theropods and they'd need to have had more saliva to swallow their food. In order to have made more saliva I believe they'd need to have had labial oral glands sitting in the inset buccal area (like lizards and snakes). I'm not saying they didn't already have oral glands which were above the roof of the mouth, but had *more* oral glands. But would this favor a cheek? I believe it would. If ornithischians didn't have cheeks the food would fall out the sides of the mouth and as would the saliva. The most beneficial use for the labial oral glands would be with cheeks and not lips. Not thick muscular cheeks as in mammals, but cheeks never the less. The ornithischians with shallow buccal area possibly just had cheeks (hypsilophodontids, dryosaurids, camptosaurids), and those with large buccal area (iguanodontids, hadrosaurids and ceratopians) would have large oral glands supported and protected by cheeks. Also, if not for cheeks or oral glands, why is there such a deep buccal area?

Con cheeks. Czerkas follows Brown & Schlaikjer (1940) and Haas (1955) in that since lizards don't have buccal muscles and dinosaurs are closer to reptiles than mammals, then it would in turn follow that dinosaurs didn't have buccal muscles. This is part of my argument about theropods not having lips. As stated above the shallowest part of the mouth, which the food has to pass through, is just behind the beak. Even though *Edmontosaurus* has the widest bill/beak, the food had to pass through the shallowest area of the mouth. This drastically cuts down the food intake area. If they didn't have cheeks they would be able to gather food from both sides of the mouth and not just the front. As the food was chopped from the sides, it would fall into the middle of the mouth, as well as a larger quantity of food.

Lips? Squamates (lizards, sphenodon and snakes) have formaina on the sides of the premaxilla, maxilla and dentary which supply veins and nerves to labial oral glands. The 'lips' cover and protect the oral glands. A third level inference extrapolates that these formaina would do the same in ornithischians; i.e. supplying veins and nerves for oral glands that would sit in the buccal area and just below the teeth. These labial oral glands would supply more saliva for the animal to swallow its food and cover the oral glands with skin similar to the lips of extant lizards. These lips would not have been the same as that of mammals but more like lizards. The lips of reptiles lack the muscular mobility that mammals have and may be the same for ornithopods. The lips may have also have helped in holding the food while being cut/eaten. Papp & Witmer, 1998, Witmer, 1998 at the SVP believed that there may have been a rhamphica that extended beyond the premaxilla and predentary in ceratopians. This has not been published on yet and I will not comment on it further. If ornithischians had cheeks, they would have covered the labial oral glands and keep food in the mouth.

In Conclusion. The traditional view of eating in ornithischians is for the beak to gather and cut the food and for the cheeks to hold the food in the mouth until cut/ground and finally swallowed. Ornithischians did not masticate their food (except for heterodontosaurs); the teeth cut or shear the food. The teeth are not held horizontal to each other as they occlude (as in mammals) but are oblique or vertical to one another. Due to small intake area of a checked animal, less food would be able to be consumed since they are gathering the food only via the 'front' of the jaw.

The food would be nipped via the premaxilla/predentary and maneuvered into the center of the mouth via the tongue which manipulates the food toward the sides of the tooth bearing area. Moving the jaws up and down and back and forth the food would be cut into smaller pieces. In more derived ornithischians the tooth bearing area was larger and extended beyond the coronid area (especially in iguanodontids, hadrosaurids, and ceratopians). The food would then be swallowed with the help of the oral glands producing salvia. But with cheeks the food wouldn't 'fall' out of the mouth and be held in the mouth for a longer period of time to be cut/sliced. The upper and lower teeth in hadrosaurids and ceratopians angle the food outward and not inward to the center of the mouth. If there were no cheeks the food would fall outside of the mouth, even if they had lips.

If they only had lips then the eating area would increase. The animal would be able to use the sides of the mouth to gather food. The animal could cut as it closed its mouth and swallow with the oral glands supplying needed saliva. The chunks of food could be cut into finer pieces with the tooth area near the coronoid holding the food in the back of the mouth in more derived ornithischians. The 'lips' would cover the oral glands and might help hold the food as it was being cut. The different sizes of the beaks may have helped in food gathering, with the shallower beaks being more precise and the wider beaks used for gathering larger quantities of food per bite. I believe cheeks were essential for eating. They would have kept the food in the mouth to be cut/sliced and having label oral glands to help make more saliva for swallowing. Just because they are reptiles does not mean they did not have cheeks. In fact there are birds that have cheeks. Greg Paul has showed that some vultures have cheeks, but that's for another article.

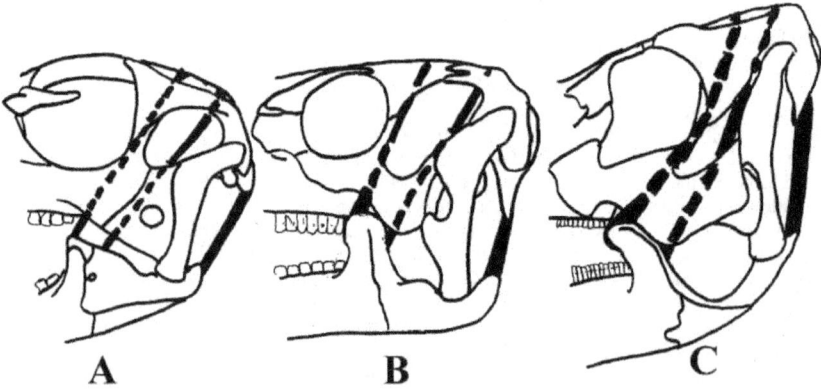

Figure 1) Jaw muscles in Ornithisichians; A) *Hypsilophodon* (after Galton, 1974); B) *Iguanodon* (after Norman, 1986); and C) *Edmontosaurus* (after Lambe, 1920),

154

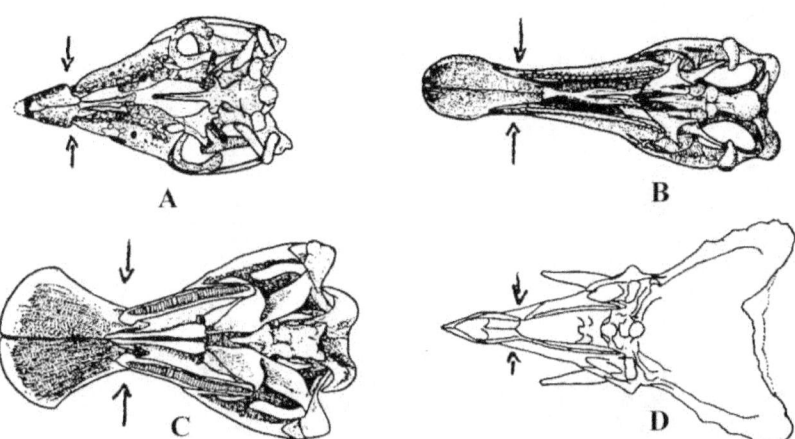

Figure 2); Bill width of ornithischians and showing the thinnest part of the mouth where food would have to go through marked with arrows; A) *Hypsilophodon foxi* (after Galton, 1974); B) *Iguanodon atherfieldensis* (after Norman, 1986); C) *Edmontosaurus regailis* (after Lambe, 1920); and D) *Triceratops (prorsus) horridus* (after Hatcher, J. B., Marsh, O. C., and Lull, R. S., 1907).

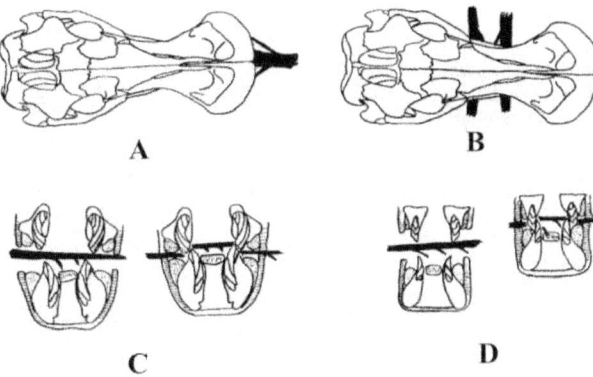

Figure 3); *Edmontosaurus* eating. A) The traditional view of food consumption from the front of the beak in dorsal view; B) from the side of the mouth; C) Transverse section showing the cutting of the plant material of D) *Edmontosaurus* and C) *Triceratops*.

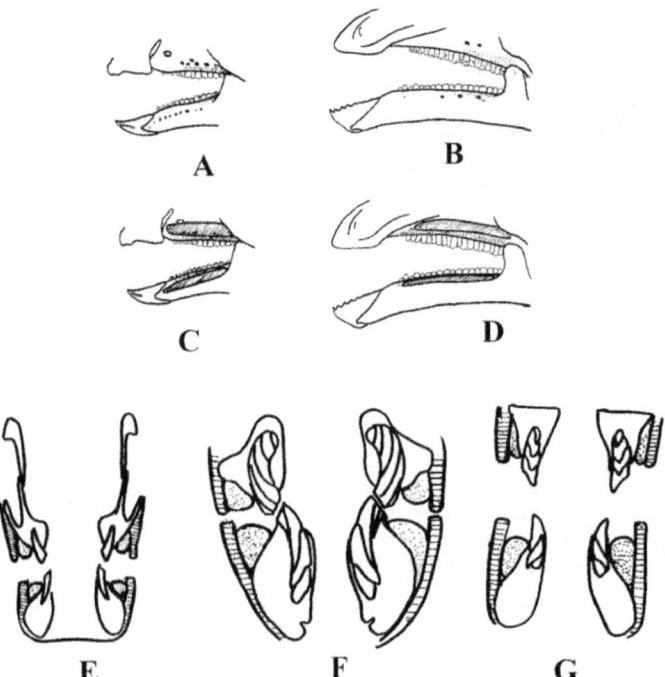

Figure 4); Maxillary and dentary region showing the foramina (in black) of A) *Hypsilophodon foxi;* B). *Iguanodon atherfieldensis*; Possible oral glands (stippled) and a layer of skin or lips (lined) of C) *Hypsilophodon*, D). *Iguanodon*; Transverse section of E); *Hypsilophodon*; F) *Iguanodon*; and G) *Triceratops*.

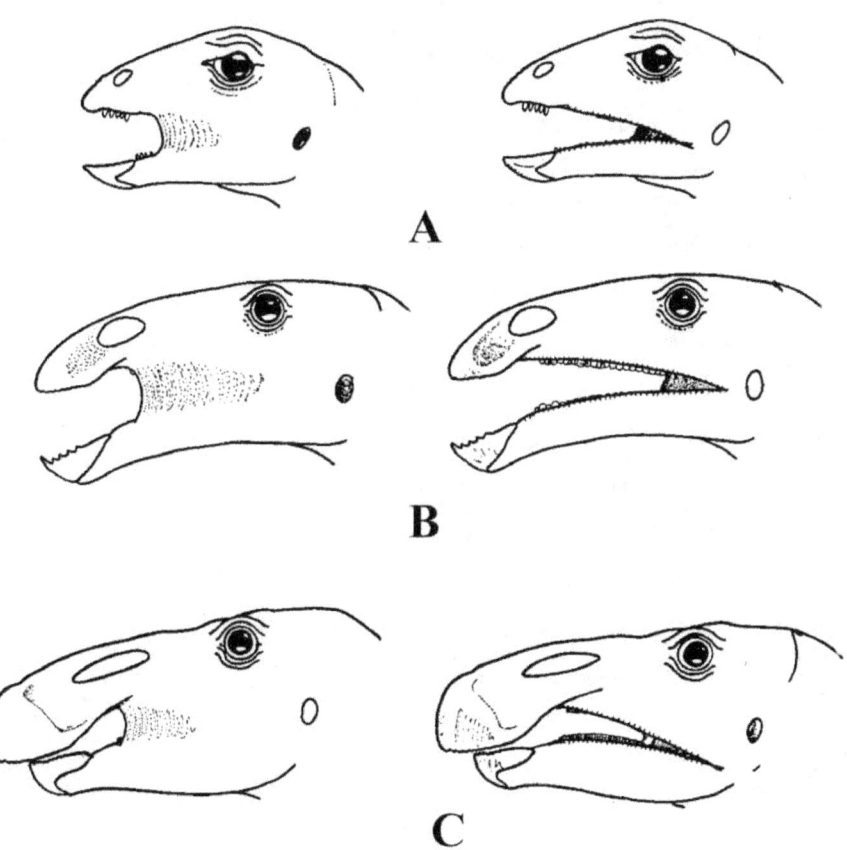

Figure 5) Restoration of the heads of ornithischians with cheeks and a newer look with lips in A) *Hypsilophodon*; B) *Iguanodon* and C) *Edmontosaurus*.

Bibliography

Czerkas, S., 1998, The lips, beaks, and cheeks of Ornithischians: Journal of Vertebrate Paleontology, v. 18, supplement to n. 3, Abstracts of papers. Fifty-eighth annual meeting, Society of Vertebrate Paleontology, Snowbird Ski and Summer Resort, Snowbird, Utah, September 30-October 3, p. 37a.

Galton, P. M., 1971, *Hypsilophodon*, the cursorial non-arboreal dinosaur: Nature, v. 231, p. 159-161.

Galton, P. M., 1971, The mode of life of *Hypsilophodon*, the supposedly arboreal ornithopod dinosaur: Lethia, v. 4, n. 4, p. 453-465.

Galton, P. M., 1972, The cheeks of ornithischian dinosaurs: Lethaia, v. 6, p. 67-89.

Galton, P. M., 1974, The Ornithischian Dinosaur Hypsilophodon from the Wealden of the Isle of Wight: Bulletin of the British Museum (Natural History), Geological Series, v. 25, n. 1, p. 3-152.

Hatcher, J. B., Marsh, O. C., and Lull, R. S., 1907, The Ceratopsia: Monographs of the United States Geological Survey, v. 49, p. 1-300.

Lambe, L. M., 1920, The Hadrosaur *Edmontosaurus* from the Upper Cretaceous of Alberta: Canadian Geological Survey Department of Mines, memoires 120, p. 1-79.

Lull, R. S., 1908, The cranial musculature and the origin of the frill in the ceratopsian dinosaurs American Journal of Science, 4th series, v. 25, p. 387-399.

Lull, R. S., and Wright, N. E., 1942, Hadrosaurian Dinosaurs of North America: The Geological Society of America, Special Paper, n. 40, p. 1-242.

Norman, D. B., 1986, On the anatomy of Iguanodon atherfieldensis (Ornithischia: Ornithopoda): Bulletin del l'Instut Royal Des Sciences Naturelles de Belgique, Sciences de la Terre, v. 56, p. 281-372.

Papp, M. J., and Witmer, L., 1998, Cheeks, beaks, or freaks: a critical appraisal of buccal soft-tissue anatomy in ornithischian dinosaurs: Journal of Vertebrate Paleontology, v. 18, supplement to n. 3, Abstracts of papers. Fifty-eighth annual meeting, Society of Vertebrate Paleontology, Snowbird Ski and Summer Resort, Snowbird, Utah, September 30-October 3, p. 69a.

Russell, L. S., 1935, Musculature and function in the Ceratopsia: Bulletin of the National Museum of Canada, v. 77, p. 39-44.

Witmer, L. M., 1998, Application of the extant phylogenetic bracket (EPB) approach to the problem of anatomical novelty in the fossil record: Journal of Vertebrate Paleontology, v. 18, supplement to n. 3, Abstracts of papers. Fifty-eighth annual meeting, Society of Vertebrate Paleontology, Snowbird Ski and Summer Resort, Snowbird, Utah, September 30-October 3, p. 87a.

www.ingramcontent.com/pod-product-compliance
Lightning Source LLC
Chambersburg PA
CBHW080657190526

45169CB00006B/2161